AF333234

NON-VALENCY INTERACTION IN ORGANIC PEROXIDES HOMOLYSIS REACTIONS

Chemistry Research and Applications

Additional books in this series can be found on Nova's website
under the Series tab.

Additional E-books in this series can be found on Nova's website
under the E-book tab.

Chemical Engineering Methods and Technology

Additional books in this series can be found on Nova's website
under the Series tab.

Additional E-books in this series can be found on Nova's website
under the E-book tab.

CHEMISTRY RESEARCH AND APPLICATIONS

NON-VALENCY INTERACTION IN ORGANIC PEROXIDES HOMOLYSIS REACTIONS

A. A. TUROVSKY, L. I. BAZYLYAK,
A. R. KYTSYA, N. A. TUROVSKY
AND G. E. ZAIKOV

Nova Science Publishers, Inc.
New York

Library of Congress Cataloging-in-Publication Data

Non-valency interaction in organic peroxides homolysis reactions / editors, A.A. Turovsky [et al.].
   p. cm.
 Includes bibliographical references and index.
 ISBN 978-1-61324-645-0 (hardcover)
 1. Valence (Theoretical chemistry) 2. Peroxides. 3. Dissociation. 4. Chemical bond. I. Turovs kyi, A. A. (Anatolii Antonovych)
   QD469.N66 2011   541'.39--dc23   2011014554

Published by Nova Science Publishers, Inc. † *New York*

CONTENTS

Introduction   **ix**

Chapter 1 **Objects and Investigation Methods** **1**

Chapter 2 **Structure of Organic Peroxides and Oxyradicals** **3**

2.1. Short Reference Data about the Structure of Organic
Peroxides *3*

2.2. Designing of the Frequencies and Forms of Normal
Oscillations for Tert.–Butylpebenzoate *5*

2.3. Calculation of IR–spectra for Tert.–
Alkylperoxymethyldialkylamines *12*

2.4. Conformational Analysis of Dialkyl– and
Dialkylsubstituted Peroxides *13*

2.5. Conformational Analysis of Peresters and others
Peroxides *24*

2.6. Information about the Structure of Oxyradicals *34*

2.7. General Electron Characteristic of Free Radicals *42*

Chapter 3 **Reactivity of Peroxy Compounds in Homolysis
Reactions** **55**

3.1. Radical Decomposition of Hydrogen Peroxide
and Dimethylperoxide *55*

3.2. Non–empirical Designing of the Hydrogen Peroxide
Homolysis *59*

3.3. Designing of the Process of Radical Decomposition
for the Molecule of Methylperacetate *61*

3.4. About Potential Energy of H_2O_2 Decomposition
with Taken into Account the Rotation Energy *63*

3.5. *Designing of the Tert.–Butylperoxymethyldimethyl-amine Molecule Homolysis*   67

3.6. *Peculiarities of Electron Structure of Peroxide Ethers*   68

3.7. *An Influence of the Substituents Nature on Homolysis of Peroxy Compounds*   69

3.8. *A Role of the Unshared Pairs into Peroxides Homolysis Reactions*   75

3.9. *About Correlation between the Enthalpie of Peroxy Compounds Formation into Equilibrium and Transition States into Reactions of their Homolytical Decomposition*   79

3.10. *About Preexponential Factors of the Homolysis Reactions of Peroxy Compounds*   81

Chapter 4   **About Reactivity of Short–Lived Oxyradicals**   **107**

4.1. *General Information about Reactivity of Free Radicals*   107

4.2. *Experimental Information about Oxyradicals*   108

4.3. *Estimation of Free Radicals Reactivity*   108

4.4. *About Energy of MO for Unshared Electron in Radicals*   111

4.5. *About Ratio between the Orbital Energy and the Energy of Electrons Repulsion for Radicals and Atoms*   119

4.6. *About Stabilization Energy of Free Radicals*   121

Chapter 5   **Investigations of kinetics and Chemical Mechanism of Organic Peroxides Thermal Decomposition**   **127**

5.1. *Reactivity and Chemical Mechanism of Tert.–Alkylperoxycyclo-pentenes–2 and Tert.–Alkylperoxytetrahydrobenzoates Thermolysis*   127

5.2. *Thermal Decomposition of Tert–alkylperoxymethylamines*   134

5.3. *About Chemical Mechanism of Nitrogen–Containing Peroxides Thermal Decomposition*   138

5.4. *Kinetics and Chemical Mechanism of Organosilicon Peroxides Thermal Decomposition*   142

5.5. *About some Aspects of the Solvent Nature Influence on Decomposition Kinetics of Organic Peroxides*   145

  *5.6. About the Forecast of Chemical Mechanism
    and Kinetic Parameters of Peroxides Homolysis* *151*

Chapter 6 **Photolysis of the Investigated Peroxides** **159**
  *6.1. Information about Photodecomposition of Organic
    Peroxides* *159*
  *6.2. Peculiarities of Peroxides Photodecomposition
    in Solid Phase* *160*
  6.3. Photodecomposition of Tert–butyl Peroxide *163*
  6.4. Photolysis of Tert–Alkylperoxymethyldialkylamines *166*
  6.5. Photodecomposition of Tert.–Butylperoxyacetals *174*
  6.6. Peresters Photodecomposition *181*
  6.7. Photolysis of Silicon–containing Peroxides *182*
  *6.8. About «Hot» Molecules and Radicals at the Peroxides
    Photodecomposition* *187*

Chapter 7 **About Reactivity of the Peroxides Decomposition
    under the Action of some Reactants** **195**
  *7.1. General Information about Reactions of Peroxides
    with Amines* *195*
  *7.2. Studies of the Reaction of Peroxides with
    1,1–diphenyl–2–pikrylhydrazyl (DPhPH)* *197*
  *7.3. Kinetics of DPhPH Interaction with Silicon–
    containing Peroxides* *203*
  *7.4. Investigation of Reaction of 1,1–diphenyl–2–
    pikrylhydrazin (DPhPH–H) with Peroxides* *207*
  *7.5. Interaction of Nitrogen–containing Peroxides
    with A–naphthylamine* *210*
  *7.6. About Interaction of Tert.–Butylperbenzoate
    with α–Naphthylamine* *214*

Concluding Remarks **219**

Afterword **223**

References **227**

Index **245**

INTRODUCTION

Chemical processes of the radical polymerization of different vinyl monomers, initiated vulcanization of rubbers, curing of the polyester resins *and ect.* require the initiators possessing by a whole complex of the definite physical–chemical properties. Organic peroxides are mainly used as the initiators. Their main advantages are the thermal stability within the definite interval of temperatures and in different media, reactivity of the primary oxy–radicals in different reactions and their initiation efficiency, non–toxicity of the final decomposition products. The complex of the above–said properties of the organic peroxides should be taken into account under solving the problem of their purposeful synthesis. Purposeful synthesis of the organic peroxides as a source of the free radicals for different chemical processes is one of the most actual problems of the peroxides chemistry. Development of the scientific principles of the indicated synthesis needs the solution of the main question – the question of the relationship between the peroxides electron and space structure and their reactivity via reactions of their homolytic decomposition. First of all, this means:

- the prognostication of the most probable chemical mechanism and kinetic parameters of the peroxides homolysis;
- the estimation of the reactivity of oxyradicals being by primary products of the homolytic decomposition of peroxides in different transformations reactions.

Intensive investigations of the peroxides homolysis kinetics lead to a great production line of the experimental data, unfortunately uncoordinated and not always comparable between themselves. The systematization and interpretation of these kinetic characteristics from the point of view the electron and space structure has a high profile under the aspect of purposeful synthesis of organic peroxides. As to this question there is only separate information in special references. Although a lot of the presented investigations have the qualitative or incidental character, nevertheless, theng r appearance testifies to the growing interest to the considered problem. Calculation of the homolysis absolute rates constants for the peroxides with the exception of the simplest ones practically is impossible; that's why it has had to use the empirical methods. Quantum–mechanical calculations of reactions profiles of the potential energy are rather by supplement to the complicated calculations of the specific cases than the evolution of general theory of chemical reactivity. One among real approaches to the solution of the touched upon questions is the upcoming direction basing on quantum–mechanical ground with the use of the reactivity indexes. Although the presented approach is empirical, however it can be considered in many cases as semi–qualitative.

Thus, the way out can be founded into determination of the relationship regularities between the electron structure of peroxides and their reactivity via the decomposition reactions. The indexes of the molecules reactivity theoretically are grounded only for the alternant hydrocarbons. An non–uniformity of the charger distribution and of the MO energy levels is typical for the most real chemical systems including the organic peroxides; that's why for every case it is necessary to carry out the separate analysis of those or else indexes applicability.

For example, systematic investigations of the organic peroxides structure by general formula $R_1R_2C(X)OOR$ $(X = O, S, OR_3, Cl, F, NR_2)$ permit to conclude, that energetically more advantageous is the *cis*–conformation, than the *trans*– one with respect to the bond $C–X$ and to the $O–O$ bond, that is the role of the non–valence atoms interactions is evident. Since the conformational transformations not touch upon indirectly the all atoms or pairs of atoms, we have layed, that these transformations are greatly caused by the non–valence atoms interactions of the central fragment. At the consideration of the relative reactivity of the compounds $R_1R_2C(X)OOR$ in homolysis reactions and, possibly, in other reactions, it is necessary to take into account the specificity of their electron structure with taken into account the space (non–valence) interactions of atoms. This implies that the applied any statical index for the interpretation of the considered peroxides reactivity should necessarily include the energy of the non–valence interactions of atoms of the central fragment. Under consideration of questions concerning to the reactivity of peroxides in homolytic decomposition reactions this assumption it was widen also on intra–atomic non–valence interactions of chain $-COOC-$.

In presented book:

- the questions concerning to the electron and space structure of the organic peroxides and their reactivity in homolysis reactions are considered;
- a role of the non–valence interactions under determination of the molecules conformation, the interpretation of their reactivity, the reasons of the activated complex stabilization, the chemical mechanism of the homolysis are explained and also the oxy–radicals reactivity is estimated.

The reactivity of peroxides not only during their thermolysis reactions but also in photolysis and decomposition reactions under the action of the nucleophiles are satisfactory interpreted from the point of view of electron structure of the central fragment with taken into account the non–valence interactions of atoms. Starting from this position *it was proposed the conditional classification of organic peroxides useful under the practical aspect, since it permits to forecast the chemical mechanism and kinetics of the peroxides homolysis reactions and in such a way facilitates to the solution of problem concerning to the purposeful synthesis of the peroxides as a sources of the free radicals.*

However, it is necessary to take into account also the fact, that the quantum–chemical calculations are enough approximate and have the relative character. In connetion with the improvement of the quantum–chemical calculations methods, and also with taken into account of the weak interactions, it can be hoped on more exact prognostication of the organic peroxides reactivity and also their purposeful synthesis.

Chapter 1

OBJECTS AND INVESTIGATION METHODS

Different peroxy compounds, which are enough various upon its structure from the one hand and are sufficient in quantitative aspect in order to do any generalizing conclusions about their structure and reactivity from the other hand, have been used as the *investigations objects*.

Criteria used for the selection of the investigation objects were as follows:

- fitting the peroxides with a wide range of the substitutes;
- relatively wide temperature interval of the peroxides thermal stability;
- practical value of the peroxy compounds, *i. e.*, the compounds, which based on their any previous investigations can be useful via the polymerization, vulcanization and *ect.* processes;
- reliability of the kinetic parameters of compounds thermal decomposition (the determination of kinetic characteristics by several methods under different conditions);
- the possibility to carry out the theoretical calculation of the investigated compounds.

Such groups of the peroxy compounds as *dialkyl peroxides, hetero–atom–derivatives of these peroxides, peroxy ethers of different acids*, some *hetero–organic peroxides* correspond to above–indicated demands.

Dialkyl peroxy compounds R_1OOR_1 *(R$_1$ = CH$_3$, C$_2$H$_5$, н. –C$_3$H$_7$, н. –C$_4$H$_9$, –C(CH$_3$)$_3$)* and other non–symmetric peroxides with *tert.*–alkyl radicals were investigated mainly only in theoretical aspect. Kinetic parameters of their thermolysis were taken from references. Among the cycloalkenyl peroxides it has been studied the *tert.*−alkylcyclopentenyl peroxides and their epoxy derivative compounds.

Tert.−alkylperoxymethyldialkylamines $(R_1)_2NCH_2OO_tAlk$, peroxysemiacetals $R_1CH(OH)OO_tAlk$, peroxyacetals $R_1CH(OR_2)OO_tAlk$ investigated by us can be referred to the heteroatom−substituted non−symmetric dialkyl peroxides.

Peresters of aliphatic, substituted aliphatic acids, unsaturated and aromatic acids have been investigated also mainly in theoretical aspect. Experimentally it was used in detail only *tert.*−alkylperoxytetrahydrobenzoates $Ar−C(O)OO−Alk$ and their epoxy derivatives.

Among the hetero−organic peroxides it was studied only some organosilicon peroxides.

Above listed investigation objects have enough wide range of the thermal stability, *namely* from 373 till 473 *K.*

Many of these objects (*namely*, *tert.*−alkylperoxycyclopentenes−2, *tert.*−alkyl-peroxysemiacetals,   *tert.*−alkylperoxyacetals, *tert.*−alkylperoxytetrahydrates and their derivatives) have been synthesized in Department of Radical Processes at the Institute of Physical−Organic Chemistry and Coal Chemistry of *AS USSR* in cooperation with the *UkrSIIPlastmass (A. E. Batog)*, organosilicon peroxides have been synthesized by *Litkovets A. K.* (the Department of Organic Chemistry at the Lviv Polytechnic Institute).

Chapter 2

STRUCTURE OF ORGANIC PEROXIDES AND OXYRADICALS

In this Chapter the questions concerning to the space and electron structure of a set of organic peroxides and free radicals being by the products of their homolytic decomposition are considered. A special attention is paid into the conformational analysis of the peroxy compounds, which has been done by different methods, but giving the same final result – the energetical advantageous of *cis*– or *hauch*–conformations of peroxides. Such non–trivial result needs a new qualitative explanation and undoubtedly should be reflected on the reactivity in different reactions.

2.1. SHORT REFERENCE DATA ABOUT THE STRUCTURE OF ORGANIC PEROXIDES

During the last years the radiographical and electron–diffraction investigations of geometry for a set of organic peroxides are carried out [1–6]. In *ref.* [7–9] it was shown that the molecules of peroxides have the nonplanar structure and the values of an interfacial angle $C\text{–}O\text{–}O\text{–}C$ are in a field $(100\text{–}140\ ^0)$.

A lot of works was dedicated to *IR*–spectral investigations of peroxides. Thus, in *ref.* [10] it was discovered two absorption bands by middle and feeble intensities for aliphatic peresters respectively in a field of $852 \pm 4\ cм^{-1}$ and $920\ cм^{-1}$. However, none among these bands is not relative to the valence vibrations of $-O-O-$ group, since in the spectra corresponding to *tert.*–butyl ethers the near bands are observed.

Davison [11] considers, that the vibration band of $-O-O-$ group in peresters is not distinguishing as a result of the peroxy group with the carbonyl interaction.

Sufficiently good into experimental and theoretical aspects it was studied the structure of hydrogen peroxide [12–15]. The *ref.* [16–19] confirm the *hauche*–configuration of H_2O_2.

Ionezawa and *Kato* [20] calculated the molecule of methylperacetate. More energetically advantageous was the *cis*–conformer (the position of carbonyl group with respect to the peroxy bond). However, authors don't explain the reasons of the stability of this conformer.

Under semi–empirical approximation the optimal parameters of the structure for methylhydroperoxide and dimethylperoxide have been found [20].

Sheppard [21] showed that the appearance of the characteristic frequencies of the group $-O-O-$ vibration in peroxides is less probable as a result of nearness of the mass groups of atoms $-O-O-$, $-C-O-$, $-C-C-$, that's why it is enough near also their vibrations. However, if the atoms *ROOR* are not in the same plane, then the transition of the vibration energy from the skeleton of the whole molecule on the bond $-O-O-$ doesn't proceed and then the characteristic frequencies for $-O-O-$ group can be appeared in a spectrum.

Kovner and others [22] carried out the theoretical calculations of vibrating groups $-COOH$ and $-COOC$. It was assumed, that the groups have a coplanar structure with a value of dihedral angle 90^0. Calculated frequencies of vibrations are in good agreement with the experimental ones.

Authors [23] showed, that enough characteristic on intensity are vibrations of the dihedral angle and angle $-OOH$, and also $-O-H-$ bond. The intensities of the remaining vibration of group $-COOH$ are less characteristic and are changed dependently on the group surroundings.

In works [23, 24] authors contented themselves with the calculation of $-COOC-$ fragment for the interpretation of peroxides by $R-C(O)OOC-R$ and $R-C(O)OOR$ type. Probably, the simplified model doesn't lead to the desired result.

So, at studies of the peroxides structure the experimental and calculated *IR*–spectroscopy are widely used. The calculations are carried out mainly with the use of the interpretation of experimental data on *IR*–spectra. However, the combination of the calculated and experimental *IR*–spectroscopy can give the useful information about conformers of the peroxides molecules and, thereby, predetermine their reactivity.

For more obligate confirmation of the synthesized cyclopentenyl peroxides and cyclo–containing peresters structure the investigations of *IR*–spectra were carried out with the aim of their structure peculiarities clarification.

2.2. DESIGNING OF THE FREQUENCIES AND FORMS OF NORMAL OSCILLATIONS FOR TERT.–BUTYLPEBENZOATE

IR–spectra of peresters have been studied in *ref.* [25]. However, the authors didn't the unequivocal assignment of the frequencies for vibration of –O–O– bond to the peresters. For example, in a case of the *tert.*–butylperbenzoate for vibrations of –O–O– bond accordingly to [23] it can be assigned the one among frequencies 828 and 860 cm^{-1}. Authors [24] carried out the calculations of vibrations of model chain $COOC$.

In *ref.* [26] there are results on calculation of the vibration for molecule of *tert.*–butylperbenzoate. We were limited by the consideration of two extreme isomers, *namely cis–* and *trans.–* Under molecule model construction the data of *ref.* [27] were used. The force constants were taken from *ref.* [24, 28–29] and were undergo to the insignificant variants. On Figure 1 the structure of *tert.*–butylperbenzoate molecule is represented.

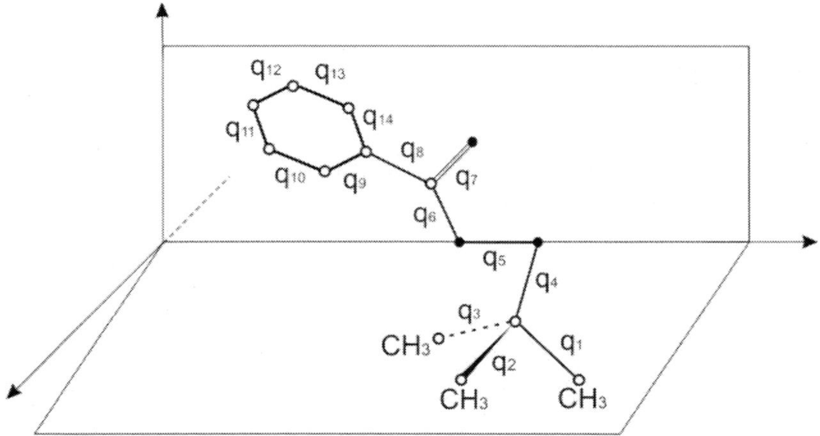

Figure 1. The structure and oscillating coordinates of *tert.*–butylperbenzoate molecule.

Table 1. Oscillation frequency of *tert.*–butylperbenzoate molecule

№	Oscillation form	Calculated, sm^{-1}		It was found, v_{sm}^{-1}
		v_{cis-}	v_{trans-}	
1	7 (C=O)	1747	1747	1755
2	1,4 (C–C)	1699	1699	1645
3	9 (C–C)	1643	1643	1618
4	8 (C–C)	1484	1484	1490
5	11 (C–C)	1337	1337	1340
6	12 (C–C)	1337	1337	1340
7	3 (C–C)	1331	1331	1315
8	10 (C–C)	1272	1272	1280
9	11 (C–C)	1272	1272	1280
10	6 (O–C)	1255	1253	1245
11	1 (C–C)	1199	1199	1178
12	4 (C–O)	1162	1162	1178
13	9,10 (CCC)	1078	1074	1078
14	9,14 (CCC)	961	–	–
15	5 (–O–O–)	944	997	933
16	5 (–O–O–)	866	859	863
17	6,7 (OCO)	801	645	830
18	2 (C–C)	732	736	750
19	11,12 (CCC)	669	725	670
20	7,8 (OCO)	578	615	–
21	6,8 (OCC)	578	615	–
22	13,14 (CCC)	548	558	–
23	10,11 (CCC)	513	525	–
24	2,4 (CCO)	475	485	475
25	1,4 (CCC)	415	414	–
26	2,3 (CCC)	415	415	–
27	1,2 (CCC)	390	390	358
28	1,3 (CCC)	366	374	345
29	3,4 (CCO)	351	343	–
30	8,14 (CCC)	293	298	306
31	5,6 (OOC)	219	212	–
32	4,5 (OOC)	171	176	–
33	8,9 (CCC)	114	98	–

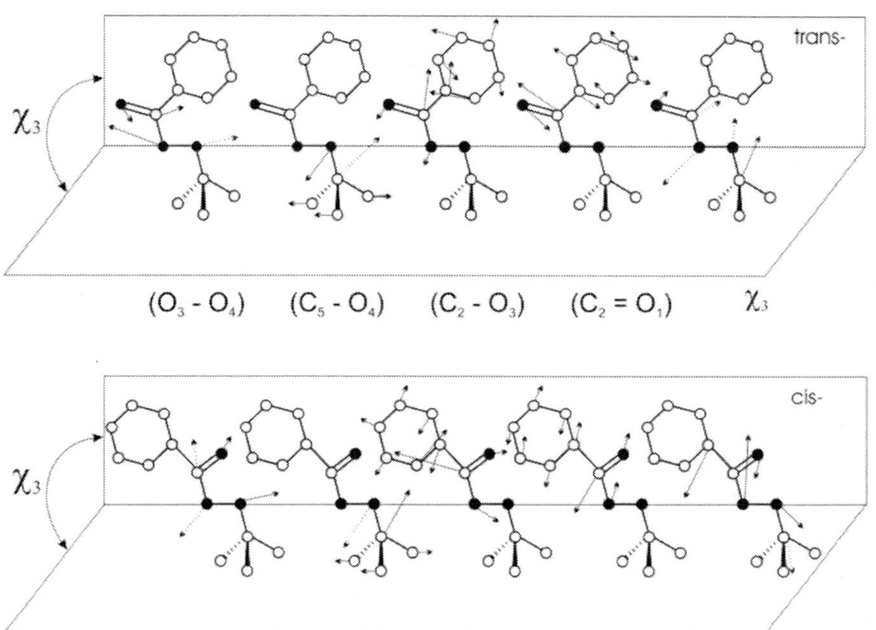

Figure 2. The structure, displacement of atoms and some frequencies of *cis–* and *trans–*conformers of *tert.–*butylperbenzoate.

As we can see from the data of Table 1, it is observed the satisfactory agreement between the experimental and calculated values of frequencies of valence and deformative vibrations.

At the transition from the *cis–* to the *trans–*conformation (the force field was not changed at this) it is observed a sharp change only of the one among valence vibrations of *–O–O–* bond.

On Figure 2 there are displacements of the atoms of *cis–* and *trans–*conformers of *tert.–*butylperbenzoate.

The maximal displacements of oxygen atoms of peroxy bonds corresponds to the frequencies 944 *sm*$^{-1}$ *(cis–)* and 997 *sm*$^{-1}$ *(trans–)*; and for the frequencies 866 *sm*$^{-1}$ *(cis–)* and 859 *sm*$^{-1}$ *(trans–)* the displacements of atoms of the peroxy bond is less, but at the same time there are strong displacements of atoms in both *C–O* bonds, that is typical for the oscillation of nonplane angles. Starting from the analysis of forms of the normal oscillations and the atoms displacements, the bands 944 *sm*$^{-1}$ and 997 *sm*$^{-1}$ were referred by us to the valence vibrations of *–O–O–* bonds, and 866 *sm*$^{-1}$ and 859 *sm*$^{-1}$ to the oscillations of the main dihedral angle in *cis–* and *trans–*conformers of *tert.–*butylperbenzoate [23]. Comparison of the calculated frequencies of normal oscillations of $-O-O-$ bond in *cis*– (944, 866 sm^{-1}) and *trans*– (997, 859 sm^{-1}) conformers with the experimental founded ones (933 and 863 sm^{-1}) permitted: *firstly*, the band of 933 sm^{-1} to refer to the oscillation of $-O-O-$ bond, and the band of 863 sm^{-1} to the oscillations of dihedral angle; *secondly*, calculated frequency of the normal oscillation of $-O-O-$ bond of *cis*– conformer is greatly better agreed with the experimental founded one that means the energetic advisability of the *cis*–conformer. Taking into account the hydrogen atoms leads to the change of the frequencies and forms of the normal oscillations of molecule *(*Figure 3 and Table 2).

Qualitatively it is confirmed the result about better agreement of the frequencies of *cis*–conformation with the experimental frequencies. Let us note that the oscillation task was solved in a pure conformation aspect and doesn't pretend on the detail interpretation of the spectrum.

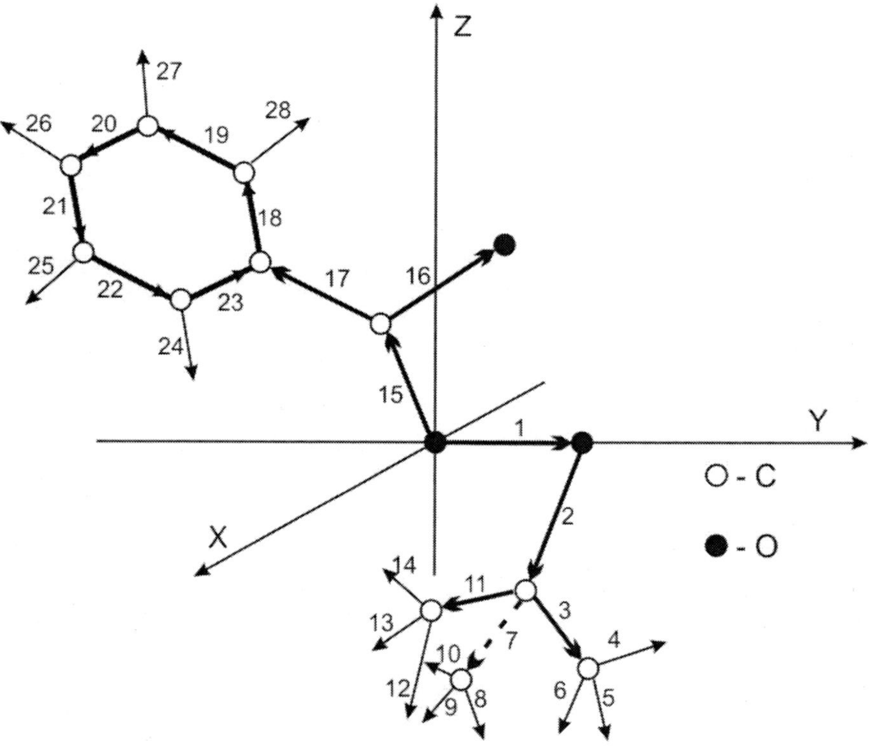

Figure 3. Structure and oscillation coordinates of the *tert.*–butylperbenzoate molecule with taken into account of the all atoms of hydrogen.

Table 2. The values of frequencies and forms of the normal oscillations for *tert.*–butylperbenzoate

№	$v_{exp.}$, sm^{-1}	v_{cis-}, см$^{-1}$	v_{trans-}, sm^{-1}	Oscillations forms
1	*2*	*3*	*4*	*5*
1	3060	3071	3071	H–C$_{ar.}$
2		3066	3066	
3		3062	3062	
4		3058	3058	
5	3030	3056	3056	
6	2985	2981	2981	
7		2981	2981	
8		2980	2980	
9	2935	2979	2979	H–C (methyl)
10		2978	2978	
11	2870	2880	2880	
12		2880	2880	
13		2880	2880	
14	1755	1769	1771	16(C=O), 15(C–O), 56(OCC), 57(OCC), 58(OCC), 59(CCC)
15	1645	1694	1694	19(C–C), 22(C–C), 23(C–C), 60(CCC), 61(CCH), 62(HCC)
16	1618	1681	1681	18(C–C), 20(C–C), 21(C–C), 22(C–C), 68(CCH), 69(CCH)
17	1490	1511	1511	23(C–C), 17(C–C), 21(C–C), 65(CCH), 71(CCH), 72(CCH), 73(CCC)
18	1475	1464	1464	21(C–C), 22(C–C), 66(CCH), 68(CCH), 69(CCH)
19	1390	1395	1395	43(HCH), 31(HCH), 37(HCH)
20	1390	1385	1385	29(HCH), 30(HCH), 42(HCH)
21	1390	1385	1385	36(HCH), 35(HCH), 41(HCH)
22	1390	1376	1376	31(HCH), 29(HCH), 30(HCH), 35(HCH), 37(HCH), 43(HCH)
23	1390	1376	1376	37(HCH), 36(HCH), 41(HCH), 43(HCH)
24	1390	1375	1375	42(HCH), 29(HCH), 30(HCH), 34(CCH), 35(HCH), 36(HCH)

Table 2. Continued

№	$\nu_{exp.}$, sm^{-1}	$\nu_{cis\to}$, см$^{-1}$	$\nu_{trans\to}$, sm^{-1}	Oscillations forms
25	1365	1366	1366	75(HCC), 61(CCH), 62(HCC), 65(CCH), 74(HCC)
26	1340	1323	1323	35(HCH), 38(CCH), 39(CCH), 42(HCH)
27	1340	1322	1322	30(HCH), 29(HCH), 31(HCH), 32(CCH), 33(CCH), 34(CCH)
28	1315	1317	1317	46(CCH), 31(HCH), 37(HCH), 43(HCH)
29	1280	1293	1293	17(C–C), 19(C–C), 61(CCH), 62(HCC)
30	1280	1293	1293	22(C–C), 17(C–C), 74(HCC), 75(CCH)
31	1245	1217	1217	68(CCH), 15(O–C), 65(CCH), 66(HCC), 69(CCH), 70(CCC), 71(HCC)
32	1188	1208	1208	65(CCH), 61(CCH), 62(HCC), 66(HCC), 71(HCC), 72(CCH), 74(HCC)
33	1178	1168	1168	2(C–O), 29(HCH), 30(HCH), 33(CCH), 34(HCH), 36(HCH), 39(CCH), 40(CCH)
34	1178	1105	1105	69(CCH), 68(CCH)
35	1078	1081	1081	7(C–C), 3(C–C), 11(C–C)
36	1078	1074	1073	3(C–C), 7(C–C), 11(C–C)
37	1060	1062	1062	15(C–O), 61(CCH), 62(HCC)
38	1023	1016	1015	20(C–C), 21(C–C), 61(CCH), 66(HCC), 71(HCC)
39	1023	962	971	63(CCC), 60(CCC), 70(CCC)
40	933	943	960	1(C–O), 32(CCH), 33(CCH), 34(CCH)
41	933	930	930	44(CCH), 38(CCH)
42	933	926	928	32(CCH), 38(CCH), 39(CCH), 44(CCH), 46(CCH)
43	933	918	918	39(CCH), 33(CCH), 34(CCH), 40(CCH), 46(CCH)
44	933	905	905	38(CCH), 32(CCH)

№	$v^{exp}.$, sm^{-1}	v_{cis-}, см$^{-1}$	v_{trans-}, sm^{-1}	Oscillations forms
45	863	865	865	76(COOC), 51(OCC), 52(OCC)
46	863	863	861	40(CCH), 39(CCH)
47	830	817	759	55(OCO), 57(OCC)
48	750	727	727	11(C–C), 3(C–C), 7(C–C)
49	687	692	683	67(CCC), 68(CCH), 69(CCH), 60(CCC)
50	670	638	652	64(CCC), 63(CCC)
51	617	609	614	33(CCH), 33(CCH), 34(CCH)
52	617	563	577	57(OCC), 56(OCC)
53	475	454	451	52(OCC), 53(OOC)
54	358	366	375	50(OCC), 49(CCC)
55	345	345	348	49(CCC)
56	345	331	331	47(CCC), 48(CCC)
57		312	310	48(CCC), 51(OCC)
58		293	293	51(OCC), 50(OCC)
59	306	202	215	59(CCC), 54(OOC), 50(OCC)
60		148	152	53(OOC), 52(OCC)
61		134	108	54(OOC), 58(CCC)

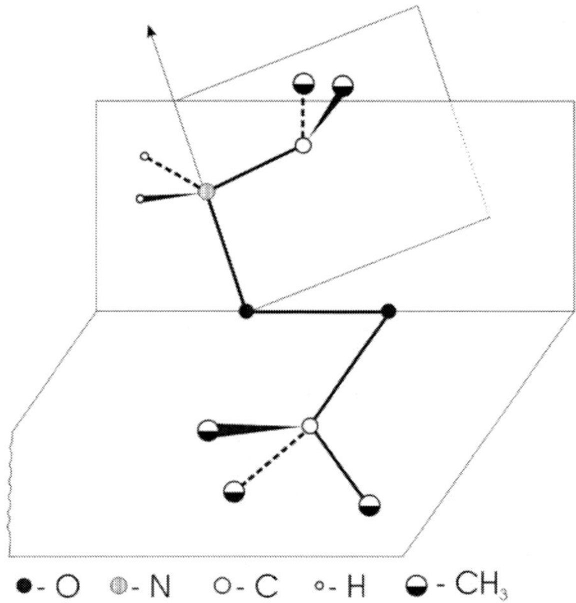

Figure 4. The structure of *tert.*–butylperoxymethyldimethylamine molecule.

Table 3. Calculated and experimentally founded oscillation frequencies of *tert.*–alkylperoxymethyldialkylamines molecules

№	Oscillation form	$(CH_3)_2NCH_2OOC(CH_3)_2$			$(C_2H_5)_2NCH_2OOC(CH_3)_3$		$(iso–C_3H_7)_2NCH_2OOC–(CH_3)_2C_2H_5$	
		Found sm^{-1}	Calculated, sm^{-1}		Found, sm^{-1}	Calc., sm^{-1}	Found sm^{-1}	Calc., sm^{-1}
			$\chi_3 = 90^0$	$\chi_3 = 175^0$		$\chi_3 = 175^0$		$\chi_3 = 175^0$
1	5,6(HCH)	1485	1638	1639	1475	1638	1470	1638
2	2,6(CH)	1370	1373	1363	1370	1367	1380	1347
3	1,7(OOC)	1252	1257	1268	1250	1265	1270	1239
4	8(O–C)	1210	1182	1189	1208	1184	1210	1180
5	1(O–C)	1075	1094	1095	1095	1094	1130	1091
6	2(C–N)	1055	1070	1067	1075	1069	1100	1080
7	3(N–C)	1035	1041	1041	1055	1040	1070	1047
8	1,5(OCH)	965	938	933	950	905	980	1025
9	7(O–O)	895	912	910	885	911	880	907
10	2,1(NCO)	630	611	622	630	612	601	587
11	3	520	587	552	520	551	520	551

2.3. CALCULATION OF IR–SPECTRA FOR TERT.– ALKYLPEROXYMETHYLDIALKYLAMINES

In *ref* [30] it was done the theoretical calculations of the *tert.*–alkylperoxymethyldialkylamines spectra. The coplanar location of the molecule fragments with the dihedral angle = 175^0 [31] was taken as the model. The force coefficients were taken from the *ref.* [22, 28, 32] and were insignificantly varied via the process. On Figure 4 it is presented the geometry of the molecule of nitrogen–containing peroxide.

As we can see from Table 3, calculated and observed frequencies are satisfactory agreed.

The oscillation frequency for the $N–C–O–O–C$ fragment represents the most interest, since it is typical for the all studied peroxides. On Figure 5 it is shown the displacements of the atoms of this fragment. It follows from this figure that что under the oscillations in a field of 800–1300 sm^{-1} the bonds – $N–C–$, $–C–O–$, $–O–O–$, and $–O–C–$ undergo the most displacement respectively. The substitutes CH_3, C_2H_5, $iso–C_3H_7$ have not an essential influence on the displacements magnitudes. This permits to consider these oscillation enough typical on purity. If to judge by frequency characteristics then the strength of the $-O-O-$ bond should be little depend on the substitutes nature. This is in good agreement with the kinetic data on thermolysis of the nitrogen–containing peroxides. It will be shown later.

The oscillations caused by the dihedral angle changes are discovered in a field of 600 sm^{-1} for the investigated compounds.

The change of χ from 90 till 175^0 for *tert.*–butylperoxymethyldimethylamine manifests itself only on the oscillation frequency of the angle.

2.4. CONFORMATIONAL ANALYSIS OF DIALKYL– AND DIALKYLSUBSTITUTED PEROXIDES

Existing interactions between the atoms in molecule considerably limits its possible configurations. For complicated molecules the atoms groups' rotation around the bonds leads to the different conformations. The potentials having the influence on turns around the bands endow the less limitations, that the potentials which determine the lengths of bonds and the angles between themselves [33].

Different on chemical structure peroxy compounds were chosen as the investigation objects. Since the experimental geometry for the complicated peroxy compounds is unknown, we have used the general principle of the complicated molecules construction from the fragments of the more simple ones.

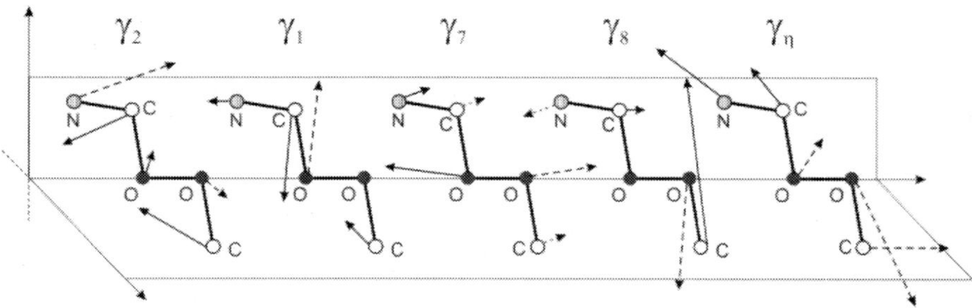

Figure 5. The atoms displacement under the oscillations of group $N-C-O-O-C-$ (the length of the vector is proportional to the displacement magnitude).

So, the models of the investigated peroxides were constructed respectively from the elements of structure of the compounds: *isobutane* [34] $(\ell_{C-C} = 1{,}525\,\overset{0}{A}, \quad \ell_{C-H} = 1{,}108\,\overset{0}{A}, \quad \ell_{C-H} = 1{,}095\,\overset{0}{A}, \quad \angle CCC = 111{,}2^0)$; *dimethylamine* [35, 36] $(\ell_{N-H} = 1{,}01\,\overset{0}{A}, \quad \ell_{C-N} = 1{,}46\,\overset{0}{A}, \quad \angle CNC = 116^0,$ $\ell_{C-H} = 1{,}093\,\overset{0}{A}, \quad \angle HNC = 112^0, \quad \angle HCH = 109{,}5^0)$; *cyclopentene* [37] $(\ell_{C1-C5} = 1{,}544\,\overset{0}{A}, \quad \ell_{C1-Cav} = 1{,}533\,\overset{0}{A}, \quad \ell_{(C-H)av} = 1{,}096\,\overset{0}{A}, \quad \angle C_1C_5C_4 = 104^0,$ $\ell_{C1-C2} = 1{,}519\,\overset{0}{A}, \quad \ell_{C-C} = 1{,}341\,\overset{0}{A}, \quad \angle C_1C_2C_3 = 111{,}2^0, \quad pucker = 28{,}8^0,$ $\angle C_2C_3C_{10} = 121{,}8^0)$; *cyclohexene* [38] $(\ell_{C1-C2} = 1{,}335\,\overset{0}{A}, \quad \ell_{C3-C4} = 1{,}515\,\overset{0}{A}, \quad \ell_{C-H} = 1{,}093\,\overset{0}{A}, \quad \angle C_1C_6C_5 = 121{,}1^0, \quad \ell_{C1-C6} = 1{,}504\,\overset{0}{A}, \quad \ell_{C4-C5} = 1{,}550\,\overset{0}{A}, \quad \angle C_2C_1C_6 = 123{,}5^0, \quad \angle C_3C_4C_5 = 111{,}0^0)$.

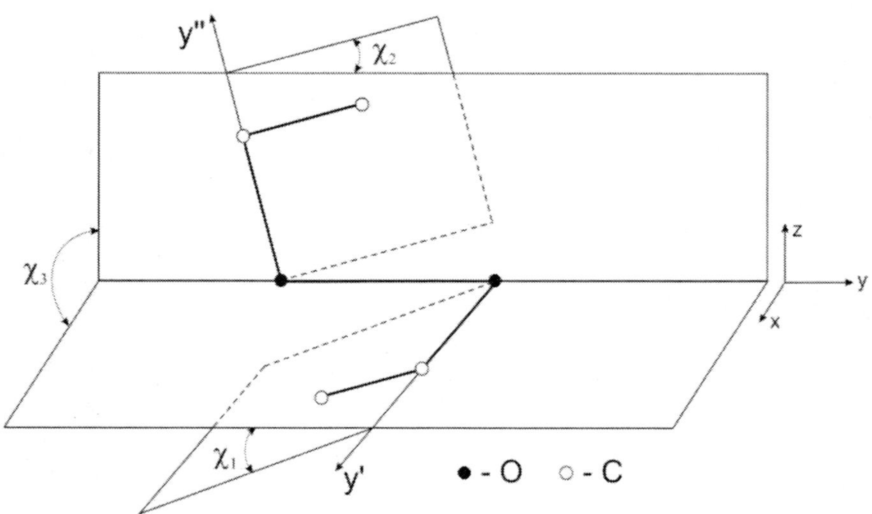

Figure 6. The indication of the dihedral angles in peroxides.

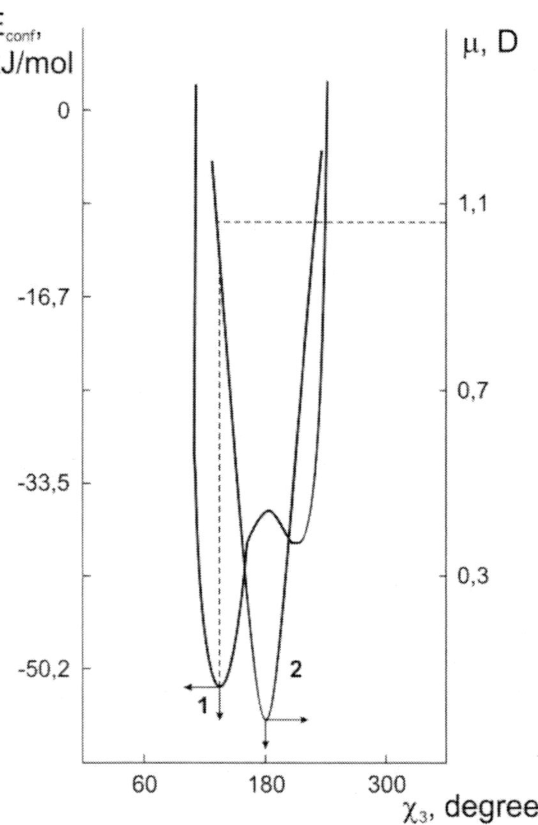

Figure 7. Dependence of the conformational energy *(1)* and value of the dipole moment *(2)* on the value of dihedral angle *(χ_3)* for molecule of *ditert.*–butylperoxide.

At the calculations of other heteroatom–containing peroxides it was used the following values of the bonds lengths: $C–F = 1,333 \overset{0}{A}$, $C–Cl = 1,767 \overset{0}{A}$, $C–S = 1,817 \overset{0}{A}$, $C=S = 1,71 \overset{0}{A}$, $C–Si = 1,870 \overset{0}{A}$, $O–Si = 1,633 \overset{0}{A}$.

Spatial arrangement of the group in complicated peroxides is characterized by the determined number of dihedral angles $(\chi_{1–3})$ *(Figure 6)*. For determination of the coordinates of atoms in molecule the *Cartesian* system was used.

(CH_3)_3COOC(CH_3)_3. On Figure 7 there are results of the conformational analysis of peroxide. On curve I it is observed two minima at 130 and 220^0. The values of the conformational energies are respectively equal to 52,04 and 42,17 *kJ*. Curve II represents a dependence of the calculated value of dipole

moment on χ_3. The value of dipole moment $1,07D$ corresponds to the energetically advantageous state of the peroxide that is in good agreement with the experimentally founded 0,96 [39]. The main dihedral angle $\chi_3 = 130^0$, $\chi_1 = 20^0$, $\chi_2 = 170^0$.

It can be seen from Figure 8, that for peroxide there are three fields (40–80^0, 140–200^0, 280–320^0), in which the deferred rotation around $-C-O-$ bonds is possible.

The value of dipole moment weakly depends on χ_1 and χ_2 *(Figure 8)*.

(CH₃)₂NCH₂OOC(CH₃)₃. Spatial structure of the nitrogen–containing peroxides was studied under approximation of both the pure electrostatic interactions [31], and with taken into account the non–valence atom–atomic interaction.

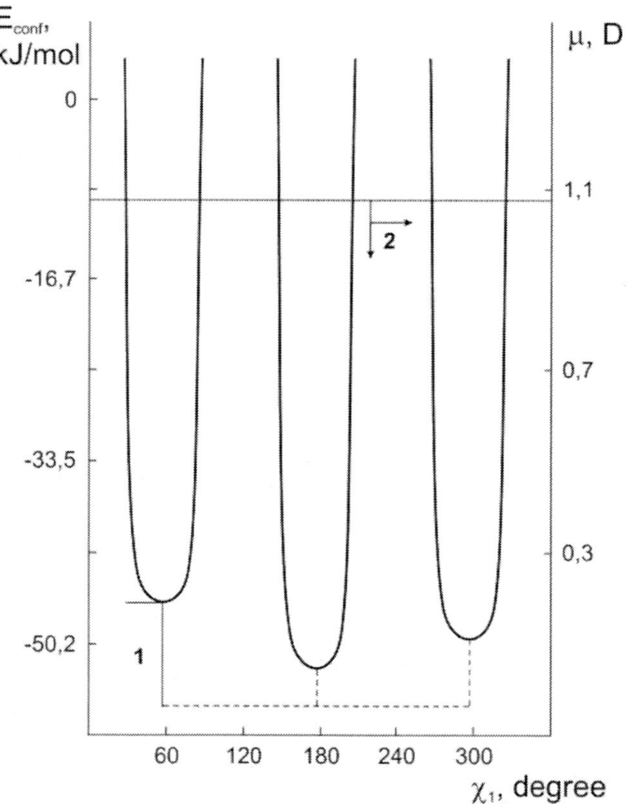

Figure 8. Dependence of the conformational energy *(1)* and the value of dipole moment *(2)* of *ditert.*–butylperoxide on value χ_1.

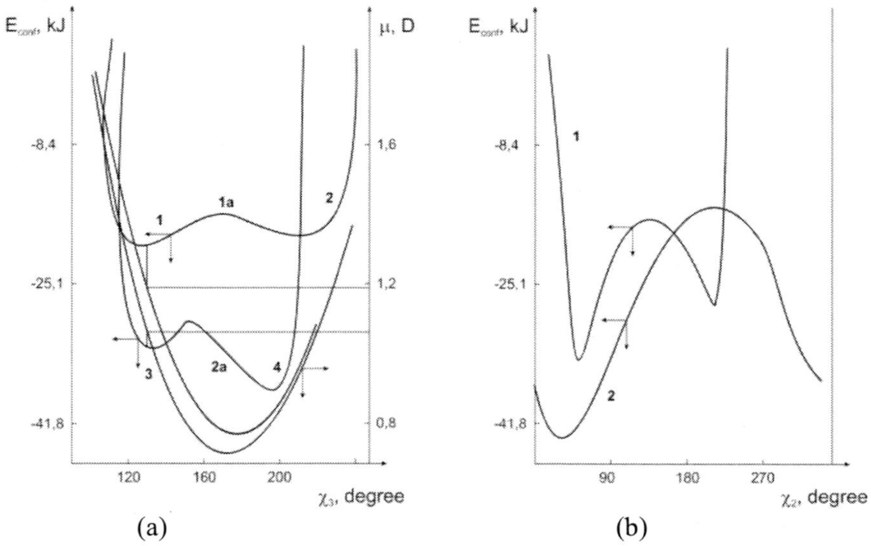

Figure 9 (a, б). Dependence of $E_{conf\phi.}$ *(Ia, IIa)* and μ on χ_3 for *cis–* *(IIa, b)* and *trans–* *(Ia, b)* conformers of *tert.–*butylperoxymethyldimethylamine *(see* Figure 9a). Dependence of $E_{conf.}$ at $\chi_3 = 130^0$ (I) and $\chi_3 = 200^0$ (II) on value χ_2 for *tert.–* butylperoxymethyldimethylamine *(see* Figure 9b).

Let us accept the minimum of the conformational atoms interactions energy as the criterion the most probable conformation. On Figure 9 there are curves of the conformational energy dependence on value of the main dihedral angle χ_3 for *cis–* and *trans–*conformers of nitrogen–containing peroxide. In a case of *trans–*conformer on curve I*a* it is observed two minima; moreover more energetically advantageous is the conformer at $\chi_3 = 130^0$. The difference of energy for conformers I and 2 at $\chi_3 = 130^0$ and $\chi_3 = 200^0$ is equal to 2,1 *kJ*. The deferred rotation around $-O-O-$ bond is possible in a diapason $\chi_3 = 110–240^0$.

At the rotation of group *(CH$_3$)$_2$NCH$_2$–* around $-C-O-$ bond it was turn that the most energetically advantageous is the *cis–*splay conformer with the value χ_2 which is equal to 30^0 *(*Figure 9, curve II*)*. In this case the rotation around the $-O-O-$ bond is described by the curve II*a* *(see* Figure 9*)* with the minima at 130 and 200^0. The second is advantageous than the first one on 10,9 *kJ*. The values of the dipole moments (curves I*b*, II*b*, *see* Figure 9) for conformers 1,3 и 2,4 respectively are equal to 1,19; 1,11 and 0,84; 0,85 *D*. Experimental value of dipole is equal to 0,95*D*.

$CH_3OOC(CH_3)_3$. (See Figure 10).

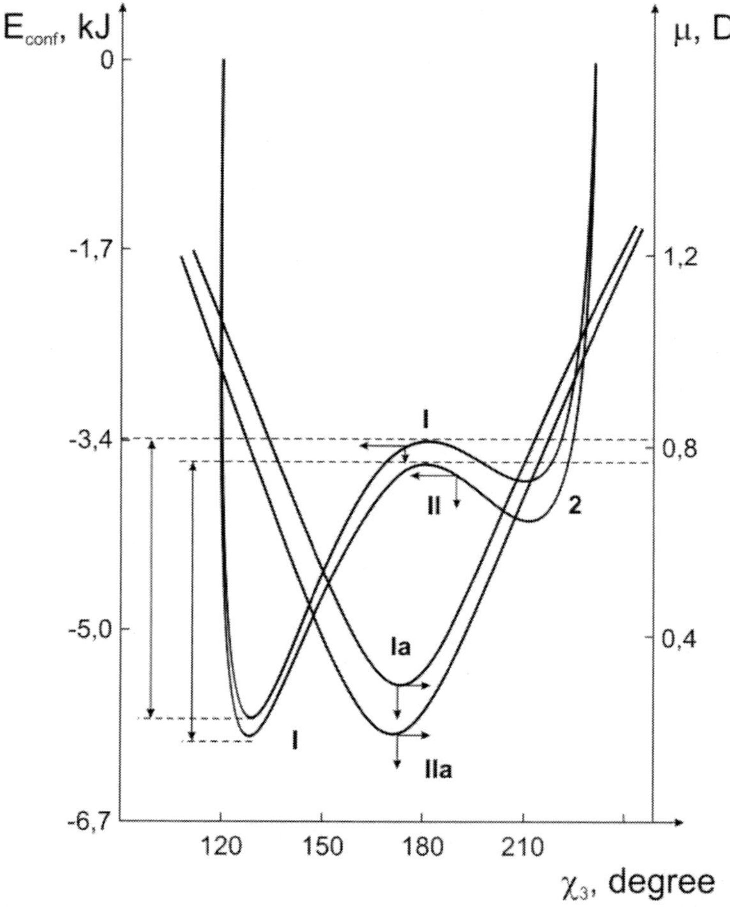

Figure 10. Dependence of the conformational energy (I, II) and dipole moment (I*a*, II*a*) for *cis–* (II) and *trans–* (I) conformers on χ_3 for molecule of *tert.–*butylperoxy methyl.

It is observed two minima on dependence curve of the conformational energy on value χ_3 in range 120–230^0; the first minimum at 130^0 is more energetically advantageous on 1,9 *kJ*. The potential barrier of the transition of conformer I into conformer II consists of 0,6 *kJ*. The value of calculated dipole moment for conformer I is equal to 0,77*D*, and for conformer II this value is equal to 0,68*D*. The rotation of methyl group around –*C–O–* bond doesn't lead to the essential changes of the conformational energy and dipole moments of conformers. Experimental value of the dipole moment is equal to 0,9 ±0,1*D*.

As we can see from Figure 11, more energetically advantageous are conformers at $\chi_3 = 130^0$. Moreover, let us note, that the *cis*–position of *C–CI* bond respectively to the peroxy one is more energetically advantageous on 6,7 *kJ*, than the *trans.*–position. The value of dipole moment for more energetically advantageous *cis*–conformer 3 *(see* Figure 11*)* is equal to 1,94*D*. For *cis*–conformer 4 the dipole moment is equal to 1,60*D*. *Cis*–conformer 3 is energetically more advantageous than the *cis*–conformer 4 on 2,5 *kJ*.

In a case of change of atom *CI* on *F* the value χ_3 is not changed. The conformation of molecule with the splay *cis*–position of *C–F* bond is some more advantageous, than the *trans*–position (1,0 *kJ*) *(see* Figure 12*)*. The value of dipole moment for more advantageous conformer is equal to 1,26 *D*.

CH₂CIOOC(CH₃)₃ and CH₂FOOC(CH₃)₃. (See Figure 11*)*.

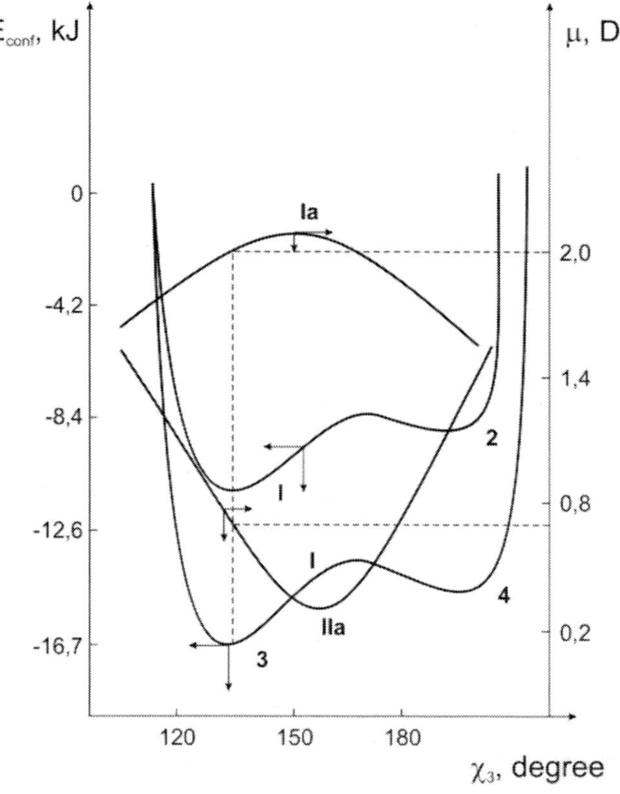

Figure 11. Dependence of $E_{conf.}$ and μ on value of χ_3 for *cis*– (I, I*a*) and *trans*– (II, II*a*) conformers of *CH₂CIOOC(CH₃)₃*.

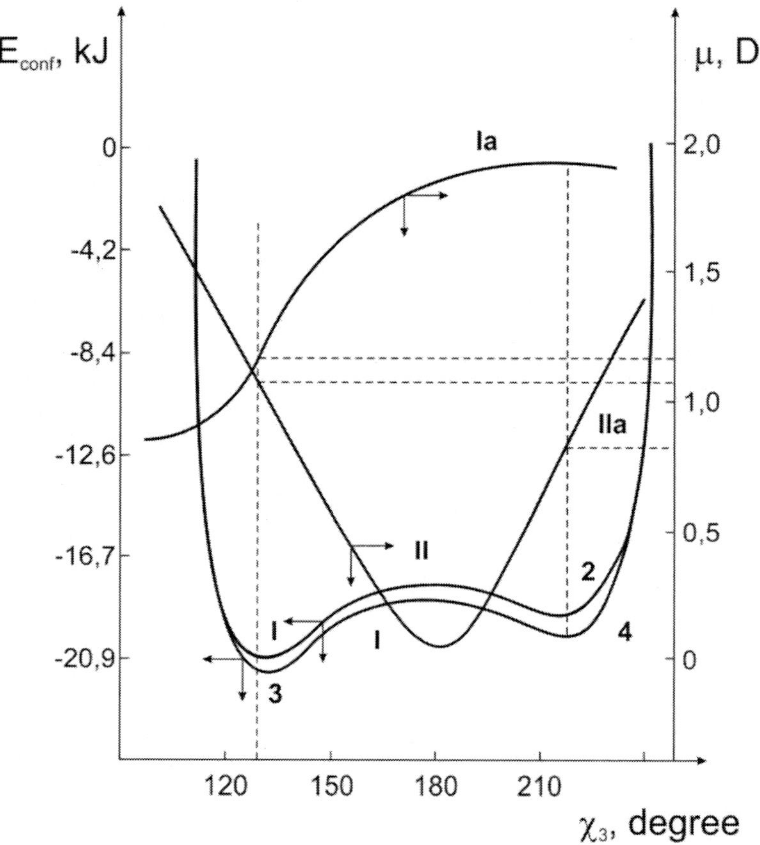

Figure 12. Dependence of $E_{conf.}$ and μ on value χ_3 for *cis*–splay $(\chi_2 = 30^0)$ *(I, Ia)* and *trans*– *(II, IIa)* conformers for *FCH₂OOC(CH₃)₃*.

$$OCH_3 \quad S–CH_3$$
$$/ \quad /$$

$H_2C–OOC(CH_3)_3$ и $H_2C–OOC(CH_3)_3$. Conformational analysis of *tert.*–butylperformal *(see* Figure 13*)* showed, that the bond $C–OCH_3$ with respect to the $–O–O–$ bond is located in *cis*–splay position and is characterized by the value of dihedral angle $\chi_2 = 55^0$, and $\chi_1 = 45^0$.

Trans–conformers are less energetically advantageous than the *cis*–conformers *(see* Figure 13, *curves* Ia, IIa)*. Calculated dipole moments of *trans*–conformers 1 and 2 are equal to 0,29*D* and 1,72*D* respectively. For *cis*–conformer the value of dipole moment is equal to 1,9*D*. Experimental value of dipole moment is equal to 2,06*D*.

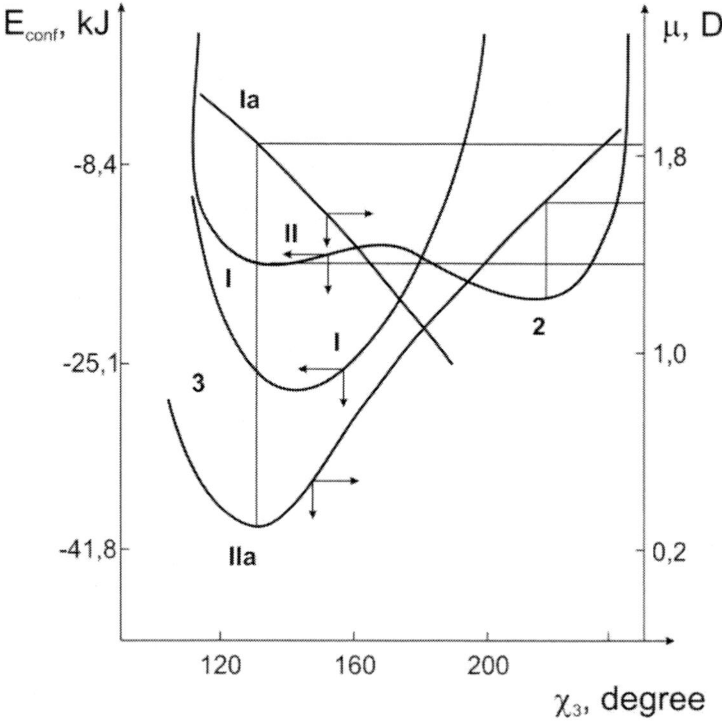

Figure 13. Dependence of $E_{conf.}$ and μ on value χ_3 for *cis*–splay $(\chi_2 = 55^0)$ *(I, Ia)* and *trans*– *(II, IIa)* conformers $H_2C(OCH_3)$–$OOC(CH_3)_3$.

In a case of the change of oxygen atom on the sulphur atom *(see* Figure 14*)* the qualitative character of the conformational energies is practically not changed. However, the angle χ_2 is equal to 45^0.

So, *cis*–splay conformation is more energetically advantageous, than the *trans*– $OOC(CH_3)_3$. On Figure 15 there are results of the conformational analysis of this peroxide. It is observed two minima for both conformers on curves of conformational energy and dipole moment dependence on value of the dihedral angle. For pseudoaxial conformer $\chi_1 = 20^0$, $\chi_2 = 170^0$ and $\chi_3 = 130^0$, and $E_{conf.} = -65,9$ *kJ* and $\mu = 1,11D$. For pseudoequatorial: $\chi_1 = 20^0$, $\chi_2 = 250^0$, $\chi_3 = 130^0$, $E_{conf.} = -74$ *kJ* and $\mu = 2,67D$.

So, the last conformer is energetically more advantageous. Experimental value of dipole moment for this peroxide is equal to $2,63 \pm 0,10D$.

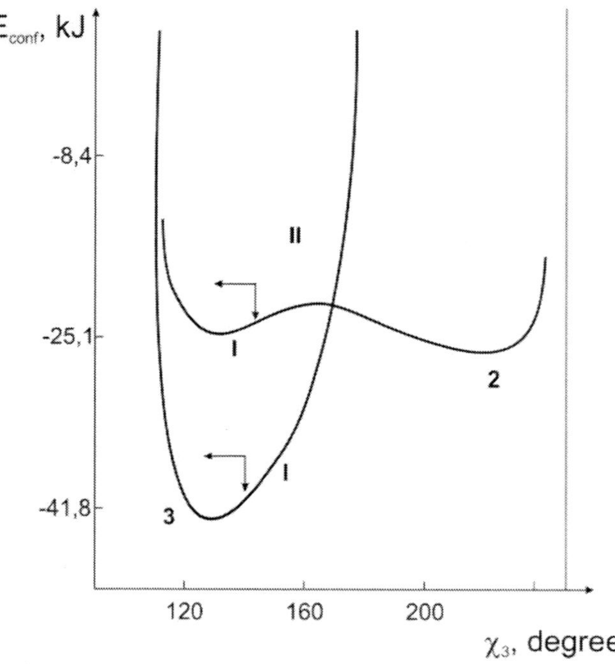

Figure 14. Dependence of $E_{conf.}$ on value of χ_3 for *cis*–splay $(\chi_2 = 45^0)$ *(I)* and *trans*–*(II)* conformers of molecule $H_2C(SCH_3)–OOC(CH_3)_3$.

Table 4. Some electron characteristics of peroxy compounds

№	Compound	Cis–conformation		Trans–conformation	
		$-E_{tot.}$, a. u.	μ, D	$-E_{tot.}$, a. u.	μ, D
1	$H_2C\overset{\diagup OCH_3}{-OO-C(CH_3)_3}$	104,7817	1,91 (2,06 exp.)	104,7688	1,71
2	$CH_3OOC(CH_3)_3$	78,7387	1,04 (0,9 exp.)	78,7284	2,17
3	$CFCH_2OOC(CH_3)_3$	94,1827	1,91	94,1444	4,18
4	$FCH_2OOC(CH_3)_3$	104,4252	1,45	104,4127	3,00
5	$(CH_3)_2NCH_2OOC(CH_3)_3$	107,5672	1,11 (~ 1 exp.)	107,5499	2,83
6	$H_2C\overset{\diagup SCH_3}{-OO-C(CH_3)_3}$	98,0770	0,95	98,0621	4,20

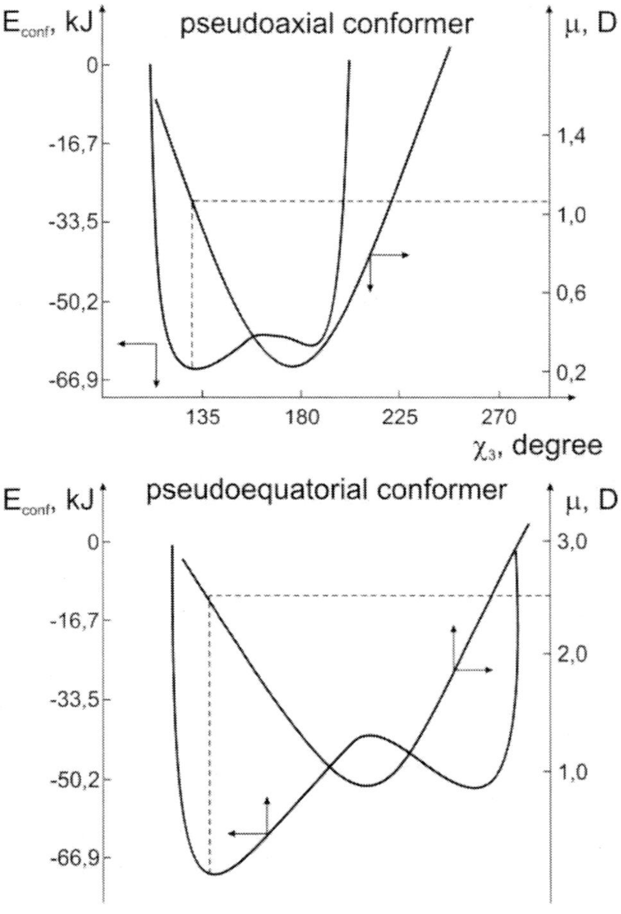

Figure 15. Dependence of $E_{conf.}$ and μ for pseudoaxial and pseudoequatorial conformers of *tert.*–butylperoxycyclopentene–2 on value χ_3.

(CH₃)₃SiOOC(CH₃)₃ and (CH₃)₃SiOOSi(CH₃)₃. The results of the conformational analysis *(CH₃)₃COOSi(CH₃)₃* are represented on Figure 16. It is necessary to note, that the experimental (1,74D) and calculated (1,51D) dipole moments are satisfactory agreed at $\chi_3 = 120^0$, that corresponds to the energetic minimum of the molecule (–56,4 kJ). For molecule *(CH₃)₃SiOOSi(CH₃)₃* the experimental dipole moment –0,9D, calculated –1,2D at $\chi_3 = 140^0$ and energetic minimum –50,2 kJ [40].

Calculation of electron structure of the above investigated and some other heteroatom–substituted peroxides also confirm that more energetically advantageous structure is the *cis*–conformation *(see* Table 4*)*.

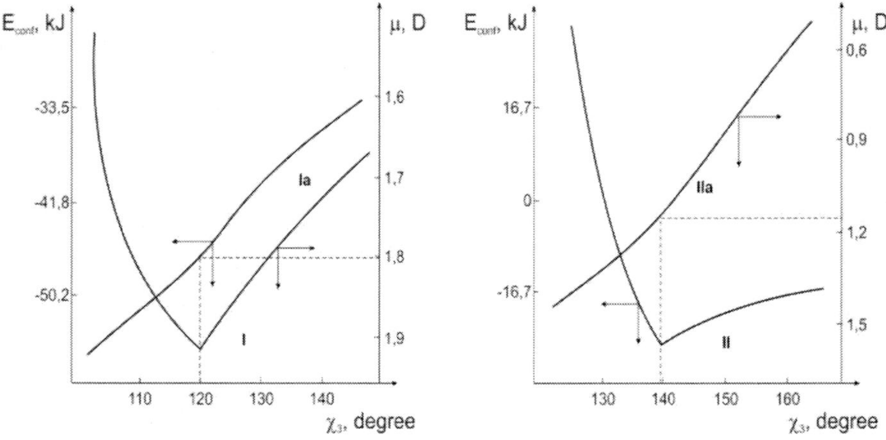

Figure 16. Dependence of E_{conf} (I) and μ (Ia) on value of χ_3 for *(CH$_3$)$_3$COOSi(CH$_3$)$_3$*. Dependence of E_{conf} (II) and μ (IIa) on value of χ_3 for *(CH$_3$)$_3$SiOOSi(CH$_3$)$_3$*.

2.5. Conformational Analysis of Peresters and others Peroxides

In *ref.* [41] it is presented the conformational analysis of peresters with the use of electrostatic model only. The charges on atoms of the molecules were calculated under approximation of *Del–Rhe* [42]. For every perester it has been studied two extreme conformers − *cis*− and *trans*−position of carbonyl group with respect to the peroxy one. Calculated dipole moments in a field of the energetic minimum on potential curve is enough greatly differed from the experimentally founded ones. This is explained by fact that the investigated method is applicable only for calculation of σ−framework of the molecule. The *Del–Rhe* method gives too much understated charges on the carbonyl oxygen; consequently the understated values of dipole moments are obtained.

Under approximation it was calculated the molecules *R–C(O)–OOC(CH$_3$)$_3$*, where *R = CH$_3$* and *i–C$_6$H$_9$* [27].

It can be seen from Table 5, that the preference should be given to the *cis*−form of the *tert.*−butylperacetate, since this form is energetically more advantageous and average dipole moment 2,66D per conformation is in good agreement with the experimentally founded 2,7 ± 0,1D.

**Table 5. A change of the electron energy *(E_tot)* for molecule of *tert.–*butylperacetate and dipole moments *(μ)* on value of the dihedral angle *(χ₃)*

№	$-E_{tot.}$, a. u.		μ, D		χ_3, degree
	cis–	trans–	cis–	trans–	
1	102,9398	–	2,02	–	70
2	–	102,80000	–	4,83	80
3	102,9889	103,0550	2,67	4,15	90
4	103,0944	–	2,57	–	95
5	103,0953	103,0871	2,60	3,62	100
6	103,0949	–	2,61	–	105
7	103,0943	–	2,62	–	110
8	103,0933	103,0905	2,66	3,71	120
9	103,0930	–	2,69	–	130
10	103,0932	103,0911	2,72	3,61	140
11	103,0918	–	2,82	–	160
12	103,0921	103,0929	2,83	5,54	180

A value of the main dihedral angle χ_3 depends on the *tert.–*butyl substitute bonds vectors location. The most advantageous is the structure of the *tert.–*butylperacetate, when the vector of bond C_9–C_{10} *(*Figure 17*)* is located in *trans–*position with respect to the $-O-O-$ bond. In this case $\chi_3 = 100^0$, and in remaining – 180^0. The location of methyl groups in *tert.–*butyl radical has not the essential influence on the value of dipole moment.

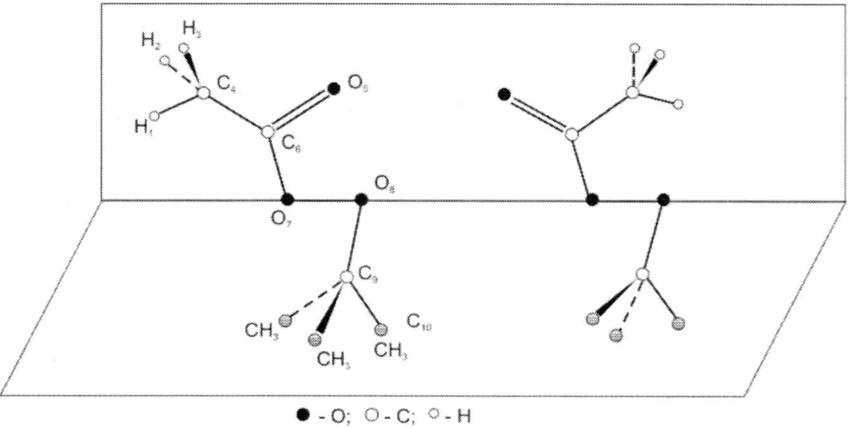

Figure 17. Model of the spatial structure of *cis–* and *trans–*conformers of *tert.–*butylperacetate.

Table 6. Electron energy and dipole moments of conformers of *tert.–*butylperoxytetrahydrobenzoate $R_1 = -C(CH_3)_3$; $R_2 = $ ⬡

№	R_2	$-C \overset{O}{\underset{OOR_2}{}}$	R_1	$-E_{tot}$, a. u.	$\mu_{calc.}$, D	$\mu_{exp.}$, D
1	Trans–	Equat.	Cis–	142,1552	3,51	–
2	Cis–	Equat.	Cis–	142,1331	2,64	–
3	Trans–	Equat.	Trans–	142,1114	3,61	–
4	Cis–	Equat.	Trans–	142,1368	2,82	–
5	Trans–	Equat.	Trans–	142,1019	3,37	–
6	Cis–	Equat.	Trans–	142,1387	2,54	2,8 ± 0,10
7	Trans–	Axial	Cis–	142,1211	3,24	–
8	Cis–	Axial	Cis–	142,1347	2,76	–

Results of the conformational analysis for *tert.–*butylperoxytetrahydrobenzoate are represented in Table 6. It can be seen from the data of this table, that energetically more advantageous is the *cis–*position of carbonyl group with respect to the peroxy group (conformers 2, 4, 6, 8), than the *trans–*position (conformers 1, 3, 5, 7).

Evidently, the radical $(CH_3)_3COO(O)C-$ occupies the axial position in cyclohexenyl, which is located in *trans–*position with respect to the carbonyl group.

For the compounds $CH_3C(O)-OOC(CH_3)_3$ and $CH_3C(O)-OO-(O)C(CH_3)$ the *trans–*conformers are also less advantageous than the *cis–* or *hauche–*forms (Table 7). The infringement of the centre planarity for compound 2 (Table 6) leads to the system energy increasing.

Summarizing the results of the peroxides conformational analysis it can be concluded that for compounds by general structure $R_1R_2C(X)OOR_3$, where X is any functional group or heteroatom, more energetically advantageous is the *cis–* or *hauche–*conformation of $C-X$ bond with respect to the peroxy bond.

Let us attempt to explain the energetic advisability of the *cis–*conformation from the point of view of the interaction of valence free atoms of the chain $\rangle C(X)-OO-C\langle-$ in peroxide.

Table 7. The values of electron energies and dipole moments for some peroxides

№	Compound	Cis–conformation		Trans–conformation	
		$-E_{tot}$, a. u.	μ, D	$-E_{tot}$, a. u.	μ, D
1	$H_3C-\overset{\displaystyle S}{\underset{\displaystyle \parallel}{C}}-OOC(CH_3)_3$	96,4595	2,94	96,4394	5,39
2	$H_3C-\overset{\displaystyle O}{\underset{\displaystyle \parallel}{C}}-OO\overset{\displaystyle O}{\underset{\displaystyle \parallel}{C}}-CH_3$	102,1248	1,30	102,1163	1,02
		102,1232 (hauche 15^0)	1,56	–	–
		102,1102 (hauche 30^0)	1,88	–	–
		102,1156 (hauche 45^0)	2,76	–	–
3	$C_6H_5-\overset{\displaystyle S}{\underset{\displaystyle \parallel}{C}}-OOC(CH_3)_3$	132,1605	2,89	132,1455	6,16

An analysis of calculations of *cis–* and *trans–*conformers of *tert.–*butylperacetate leads to the conclusion that in *cis–*conformer there is a range of *MO*, in which the atomic components of the carbonyl oxygen $O\gamma$ and O_β of peroxy bridge are overlapped, *i. e.* interact. This is illustrated on Figure 18*a*. As we can see from Figure 18*b*, for *trans–*conformer the similar phenomenon is not observed.

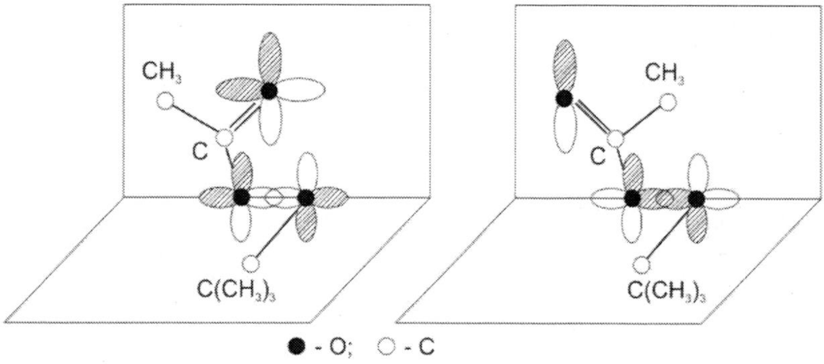

Figure 18. *MO* of *cis–* and *trans–*conformers of *tert.–*butylperacetate.

Supposing, that the interacting molecules are located still far from one from another, it can be used the perturbation theory for the estimation of perturbation energy *(ΔE)*. In this case [43] *ΔE* is represented as follows

$$\Delta E = \varepsilon_0 + \varepsilon_I + \varepsilon_D + \varepsilon_{st} \tag{1}$$

where ε_0 is interaction energy by the zeroth order (coulomb and exchange energies), ε_I is inductive component of the interaction energy, ε_{st} is energy of charge transfer related with the cross excitations, ε_D is dispersive component determined by double excited configurations.

Coulomb interaction is of the essence far apart. As far as reagents converging, the charge transfer takes place, donor and acceptor neutralize each other and the role of coulomb interaction is decreased. The energy of exchange interaction characterizes the repulsion of two reagents and, in particular, determines the multiplet separation at reactions of systems with the odd number of electrons. In general case this value is proportional to S^2, and its endowment is relatively less with respect to other terms. An interaction with the charge transfer ε_{st} is usually more essential, than the inductive ε_I. At last, the dispersive energy ε_D can play an important role at interaction between absolutely non–polar molecules.

Full calculation of *ΔE* (combinative index of the reactivity) is conjugated with the existing complications by calculating type and by the following interpretation of the obtained results [43]. That is why in references the attempts taking into account of some terms of the combinative index are appeared. The one among successful approaches has been developed by *Kloppmann* [43]. He included only the term of the zeroth order and once excited configurations with the charge transfer [43] into wave function ψ.

Let us consider from these positions an interaction of peroxides reactive centre atoms.

In a case of *cis*–conformer the electron transfer from the field of peroxy bond into field of carbonyl takes place, since the charge of oxygen of peroxy bridge for *cis*–form is some less, than for *trans.*–form, and formally the transformation process of the *trans*–conformer of peroxide molecule in *cis*–form can be considered as the process of the internal perturbance.

The equations of perturbances general method remain by suitable, but now include the electrons transfer from the one field to another in the same molecular orbital, *i. e.* both centers (donor and acceptor) connected by general molecular orbital and the perturbance has dealings with only change of electrons energy of every occupied *MO* in view of the «self–perturbance».

Equation of *HMO* for such «degenerate» case is as follows [43].

$$\Delta E = -\frac{q_r q_s}{\varepsilon R_{rs}} + 2\sum_m c_r^m c_s^m \Delta \beta_{rs} \tag{2}$$

where ΔE is energy of the interaction of atoms r and s, q_r and q_s are charges on atoms r and s, R_{rs} is the distance between atoms, c_r^m and c_s^m are coefficients of *AO* atoms r and s in $m - MO$ of molecule, $\Delta \beta_{rs}$ is the difference of energy of resonance integrals of interaction of atoms r and s, ε is the dielectric constant of medium. The first term of *eq.* (2) takes into account the electrostatic interaction of atoms; the second term characterizes the orbital interaction both valence connected and valence free atoms.

As a result of the system «self–perturbance» the corresponding resonance integrals of the non–connected atoms are changed from zero in *trans*–form to definite values in *cis*–form.

Orbital interaction of valence non–connected atoms of reactive center of peroxide in molecule we will be called by the energy of non–valence interactions. This energy is a measure of strength of valence free between themselves atoms and its value determines the conformational stability of peroxides.

From *ref.* [44] it is known a number of facts that some chemical compounds (aldehydes, unsaturated hydrocarbons, chloroanhydrides of acids, organic acids and *ect.*), in which group $C=O$ or $C=C$, $C=N$ is screened with respect to the atom of hydrogen, methyl group, chlorine atom and *et cetera* exist in *cis*–conformations.

Conformations of listed compounds are surely experimentally proved with the use of such modern methods as *NMR–*, *IR–*, *Raman*–spectroscopy, spectra of electrons diffraction and *ect.*

There are different explanations for this effect.

Thus, for the formic acid, its methyl and vinyl ethers the data are explained by the repulsion of p and π–electron cloud lying in *cis*–position. For the oxyme formaldehyde the explanation consists in the repulsion of p–p electrons.

However, in a case of propionic aldehyde and butene–1 it is admitted the presence of magnetic interaction of dipoles C–CH_3 and $C=O$.

Generally, the author [45] thinks that the interaction between carbonyl atom of oxygen and located nearby it into the intermolecular volume other polar atom (oxygen, halogen, nitrogen) leads to the visualization of the field effect. The nature of this effect is unknown. But it is supposed that it can be considered as simple electrostatic interaction leading to the change of both atoms polarities. Electrostatic interaction of atoms enough strongly can be appeared at the distances considerably greater than the sum of the *Van der Waals* radiuses.

However, in the most cases the effect visualization concerns to the atoms of halogen and oxygen located on the distances sufficiently less in order to assume the possibility of the electron unshared pairs repulsion.

Thus, the electrostatic interaction hardly can explain the real effect.

An analysis both of the experimental data and the theoretical ones upon *IR*–spectra and dipole moments for peroxides leads to the conclusion, that the *cis*–conformation is energetically more advantageous, than the *trans*–form. Conformational analysis of peroxides carried out under the approximation of atom–atom potential and *CNDO* method (partial neglect by differential overlapping) gives the same qualitative results, *i. e.* the *cis*–conformer is energetically more advantageous.

However, the potential barriers of rotation calculated with the use of these methods around the $-O-O-$ bond are greatly differed. Under approximation of atom–atom potential with taking into account the electrostatic interactions for compounds $(CH_3)_2N{-}CH_2OOC(CH_3)_3$ (*I*), $CH_2ClOOC(CH_2)_3$ (*II*), $CH_2FOOC(CH_3)_3$ (*III*), $H_2C{\overset{\diagup OCH_3}{\diagdown OOC(CH_3)_3}}$ (*IV*), $H_2C{\overset{\diagup SCH_3}{\diagdown OOC(CH_3)_3}}$ (*V*) the potential barriers for transition from *cis*– into *trans*–form are respectively equal to 16,7 *kJ*, 6,7 *kJ*, 0,7 *kJ*, 8,8 *kJ*, 15,1 *kJ*, whereas for *I, III, IV*, and *V* under approximation of *CNDO* method these barriers are respectively equal to: 45 *kJ*, 32,8 *kJ*, 34,1 *kJ*, 39,4 *kJ*. In general, it is observed some parallelism between the values of barriers, obtained with the use of both methods with the exception for compound *III*.

Under approximation of the *CNDO* method for molecules $CH_3C(O){-}OOC(CH_3)_3$ (*VI*), $C_6H_5C(O){-}OOC(CH_3)_3$ (*VII*), $CH_3C(S){-}OOC(CH_3)_3$ (*VIII*) and $CH_3C(O){-}OOC(O){-}CH_3$ (*IX*), the barriers are equal to: 21,5 *kJ*, 39,3 *kJ*, 52,5 *kJ*, and 19,7 *kJ* respectively. In accordance with [46] the atom–atom potential always gives the understated values of barriers, whereas the *NDO* methods give the barriers nearer to the experimental. Besides, the rotation barrier under approximation of completely localized approach it is obtained as the effect of *bi*– and *tri*– partial overlaps of untied atoms. Via the framework of *NDO* methods the rotation barrier is represented as the localization effect.

This fact illustrates the phenomenological character of semi–empirical methods. In the latest methods the overlapping effects are neglected, but implicitly they are taken into account with the help of the matrix elements of *Fok's* operator for atomic orbitals on different atoms. These values are calculated in accordance with the use of the formulas by *Walsberger–Helmholtz* type

$$\beta_{\mu\nu} = S_{\mu\nu}\left(\beta_{\mu}^{0} + \beta_{\nu}^{0}\right)/2.$$

As to the question of ratio *cis–* and *trans–*conformers for peroxides, it is necessary to mark following следующее. For example, the rotation barrier around the $-O-O-$ bond for peroxide $CH_3-C\overset{\displaystyle\nearrow S}{\underset{}{-\!\!-\!\!-}}OOC(CH_3)_3$ is equal to 52,5 *kJ*, and the ratio of conformers at 300 ^{0}K is equal to $\dfrac{N_1}{N_2} = 8,4\cdot 10^{-10}$, *i. e.* in fact the peroxide exists exceptionally in *cis–*form.

For compound $CH_3–C(O)–OOC(CH_3)_3$ the barrier is some less – 21,5 *kJ.* The ratio N_1/N_2 at 300 ^{0}K is equal $2\cdot 10^{-4}$, *i. e.* the molecule exists only in *cis–*conformation.

However, the conformation for compounds by general formula $R\overset{\displaystyle\nearrow X}{\underset{\displaystyle R\nwarrow}{-\!\!\!C-\!\!-}}O\text{-}O\text{-}R$ can be observed, since these molecules can exist in some *cis–*conformations caused by the vibrations of the main dihedral angle.

However, the transformation barriers of these conformers are insignificant. Evidently, it can be considered that their reactivity via homolysis reactions is approximately the same.

Thus, summarizing the above–said, it is necessary to mark following.

The evidence of the peroxides by general formula $R_1\overset{\displaystyle\nearrow X}{\underset{\displaystyle R_2\nearrow}{-\!\!\!C-\!\!-}}O\text{-}O\text{-}R_3$

(X = O, F, CI, NR₂) existence in *cis–*conformation is:

- *firstly,* from references it is well–known a series of reliable experimental facts about more stable *cis–*conformations for a number of organic compounds (aldehydes, acids, ethers *and other* compounds), containing the carbonyl group;
- *secondly,* the comparison of experimental and theoretical *IR–*spectra for *tert.–*butylperbenzoate leads to the conclusion about greater stability of its *cis–*form;

- *thirdly*, the conformational analysis under approximations of atom–atom potential with taking into account of the electrostatic interactions and *CNDO* method for a number of peroxides gives the possibility to conclude about the preferring of the *cis*–conformations;
- *fourthly*, the independence of dipole moments on temperature is evident experimental proof of the peroxides existence in the same, more energetically advantageous *cis*–conformation;
- *fifthly*, the identity of *IR*–spectra for peroxide of *tert.*–butylperbenzoate in liquid and frozen states evidently testifies that for the compound is typical only one conformer, *i. e. cis*–form *(see* Figure 19*a,* 19*b)*;
- *sixthly*, the explanation of kinetics and chemical mechanism of the peroxides homolysis from the point of view of their *cis*–structure is the main argument in favour of the correct approach to the studies of organic peroxides structure.

As to the theoretical explanation of the existence of more energetically advantageous *cis*–conformation for molecules of peroxides, it is necessary to mark the following.

Firstly, full conformational analysis of peroxide molecule under approximation of the atom–atom potential with taking into account of the electrostatic interactions in general correctly predicts the conformations, which permit to conclude about the energetic advantageous of *cis*–conformer. The barriers of rotation around the $-O-O-$ bond calculated under approximation of atom–atom potential and *CNDO* methods are symbate. Although in the first case they considerably understated, most likely due to the ignorant calculation of the repulsion forces.

Secondly, taking into account of the quantum–chemical interactions (orbital interaction) even for atoms of the reactive centre explains the stabilization of *cis*–conformations of peroxides.

So, total orbital interaction of valence free atoms of the reactive center with taking into account of the orbital interaction of atoms of $O-O$ bond can be substantiated for its using as the index of the peroxides reactivity in different reactions.

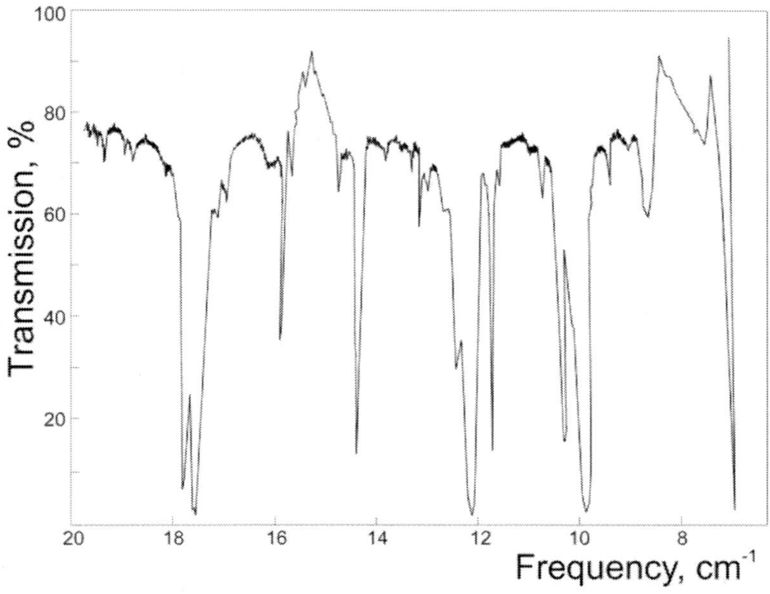

Figure 19a. IR–spectra of *tert.*–butylperbenzoate in CCl_4 at temperature 298 K.

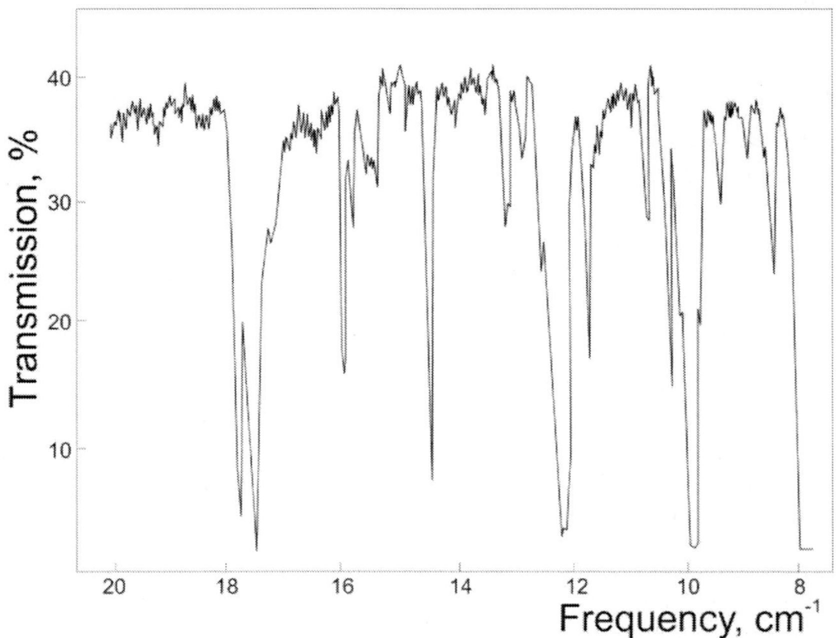

Figure 19b. *IR*–spectra of *tert.*–butylperbenzoate in CCl_4 at temperature 243 K.

2.6. INFORMATION ABOUT
THE STRUCTURE OF OXYRADICALS

Geometry of the short–lived alkoxy–radicals was the object only for limited number of the investigations via the framework of the experimental studies. Up to the present time it was experimentally determined only the length of a bond in radical OH ($\ell_{O-H} = 0,971\overset{0}{A}$) [47]. The distance of $O–H$ equal to $0,961\overset{0}{A}$, calculated under the approximation of $CNDO/2$ method, is some shorter, but is near to the experimental value, than $1,026\overset{0}{A}$ [48] $INDO/2$ or $1,033\overset{0}{A}$ $CNDO$ [49]. At the same time, in a case of CH_3O $\ell_{C-O} = 1,305\overset{0}{A}$ ($MCNDO/2$) is too much short in comparison with the $\ell_{C-O} = 1,44\overset{0}{A}$ (ab initio) [50], $1,36\overset{0}{A}$ [49] (CNDO), $1,35\overset{0}{A}$ [49] (INDO/2).

As the calculations with the use of $CNDO$ or $INDO/2$ (full neglecting by the differential overlapping) methods usually give the short lengths of bonds due to the deficiency of the taking into account of the electron–electron repulsion [51], the length of the bond $C–O$ in CH_3O ($\ell_{C-O} = 1,305\overset{0}{A}$) seems to be unappreciated in connection with the overestimation of the angle HCO ($114,8\,^0$), i. e. if more accurately founded value of the angle HCO consists of 109^0 (ab initio), then the predicted length of a bond $C–H$ in accordance with the $MCNDO$ method is acceptable. The electron repulsions and especially between the atoms H and unshared pair of oxygen in CH_3O should, probably, makes the bond more, than $1,305\overset{0}{A}$. Under this sense the calculated value of $\ell_{C-O} = 1,302\overset{0}{A}$ in C_2H_5O is also too much understated in view of too much big angle CCO (INDO/2) in comparison with the 108^0 (CNDO) [51].

The ionization potentials, estimated by different semi–empirical methods (MCNDO/2, INDO/2, CNDO) among the same range of the substitutes for peroxyradicals are less, than for the alkoxyradicals. As to the dipole moments reflecting the electron distribution in molecule, the results of the $MCNDO$ methods give relatively acceptable values for peroxides and radicals almost with the same accuracy as same as other semi–empirical methods [51].

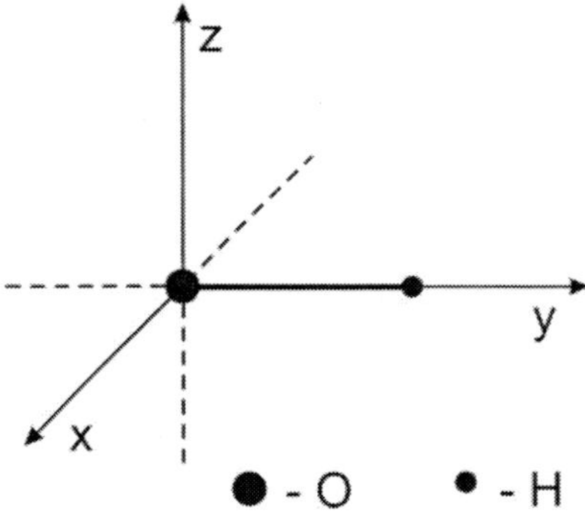

Figure 20. Geometry of radical OH $(\ell_{O-H} = 1{,}040\,\overset{0}{A})$.

2.6.1. Hydroxyl Radical

For the simplicity of the statement we will be consider the electron structure of the radical into concrete system of the coordinates (Figure 20).

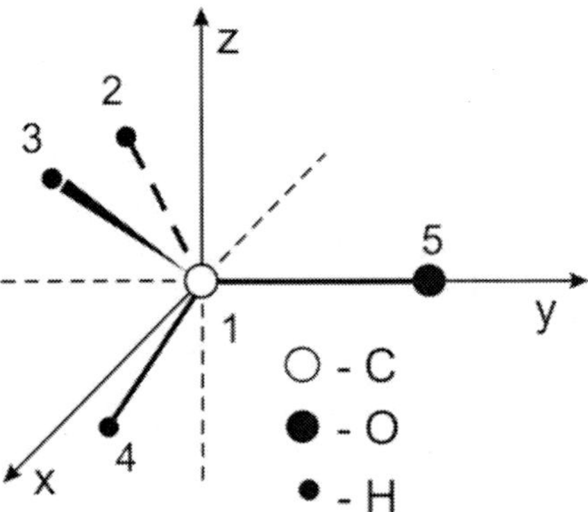

Figure 21. Geometry of radical CH_3O.

In radical there are 7 MO, on which the 7 electrons are located (4 with α–spin, 3 with β–spin). The properties of radicals in many respects are determined by the unpaired electron (its energy of MO, electron density, spin and ect.).

Analysis of the own vectors of own values of MO for hydroxyl radical for α and β spin gives the possibility to locate the seven electrons on one–electron

MO and to determine, which *MO* gives the main endowment into *O–H* bond, and on which *MO* the unpaired electron is located.

The unshared *s*–pair of electrons occupies the first *MO (α* and *β)*. This pair gives its endowment into the loosening of the *O–H* bond. The first *MO* in the loosening one per its nature. The second *MO* is occupied by electrons P_y of oxygen and *s*–electron of hydrogen and is the connecting one. The third and fourth *MO* are practically localized. The third αMO–P_z, and the third β–P_x. The fourth *MO* is bounding. The fourth αMO–P_x occupied and the fourth βMO–P_x is free. So, the fourth αMO and the third βMO occupied by the electrons which form the unshared pair. On the third αMO there is free unpaired electron. Free fourth *MO* corresponds to this orbital; the fifth *MO* is loosening and is responsible for the *O–H* bond.

So, the unpaired electron occupies not the boundary *MO*, but the *MO* with lower energy.

2.6.2. Oxymethyl Radical

Parameters of the optimal geometry of radical are as follows:

$$\ell_{C-O} = 1{,}36 \overset{0}{A}; \; \ell_{C-Hav} = 1{,}12 \overset{0}{A}; \; 415 = 108^0 17; \; 215 = 109^0 28; \; 214 = 108^0 54; \; 213 = 110^0 03.$$

Spin density on atoms is represented in Table 8.

Maximal endowment into the spin density of radical has the electron, lying in *p*–orbital of oxygen (0,9616).

In oxymethyl radical there are 22 one–electron *MO*, on which the 13 electrons substituted – 7 with the *α*–spin and 6 with the *β*–spin. On the first *MO (α* and *β)* there are *s*–electrons of the oxygen atom and electrons of carbon. These *MO* are connecting upon nature. The following three *MO (α* and *β)* correspond to the *C–H* bonds. The fifth *MO*–connecting has the maximal endowment into the *C–O* bond formation.

Table 8. The values of spin densities *(ρ)* on atoms CH_3O

Atom	C_1	H_2	H_3	H_4	O_5
ρ	0,1118	0,0252	0,0252	0,0997	0,9616

Boundary MO (α and β) occupied by the electrons of the unshared pair, and on the sixth α–orbital there is the unpaired electron. Boundary free seventh β–MO corresponds to this orbital. Unpaired electron is not localized on the sixth α–MO. Its density partially delocalized also on the third α–MO. Unpaired electron with P_z–orbital of carbon and s–orbitals of hydrogen forms the pseudo–π–system.

In Table 9 there are occupations of AO of radical CH_3O. It is clear seen from the presented Table 9 that the unpaired electron mainly is localizated on the P_z–orbital of oxygen. Judging upon the occupations the atoms of hydrogen in CH_3O radical are unequal as a result of the interaction of one proton with the unpaired electron. Constants for CH_3O are equal to: $-12,49$; $13,63$; $13,63$; $53,84$; $15,23$ for H, H, H, C and O respectively.

So, taking into account the diffusion spin and electron density upon the orbitals it can be distinguished the unshared pair P_x, unpaired electron P_z and orbital P_y of oxygen, forming the bond C–O with the AO of the carbon.

Table 9. The values of occupations of AO (Q) for radical CH_3O

Atom	Atomic orbital	Q_α	Q_β	$Q_{\alpha+\beta}$
C	S	0,5196	0,5348	1,0544
	P_x	0,4588	0,4708	0,9296
	P_y	0,3998	0,4312	0,8308
	P_z	0,4460	0,4991	0,9451
H	S	0,5291	0,5039	1,0330
H	S	0,5291	0,5039	1,0330
H	S	0,5291	0,4484	0,9966
O	S	0,9196	0,9024	1,8220
	P_x	0,9797	0,9784	1,9581
	P_y	0,6883	0,6741	1,3624
	P_z	0,9820	0,0531	1,0350

2.6.3. Spatial and Electron Structures of Oxyacyl Radical

Under the *CNDO* method approximation of the extremum direct search it was founded the optimal geometry of the nuclear skeleton for radical $CH_3C\overset{/\!/O}{\underset{\diagdown O^*}{}}$. The radical's parameters are represented on Figure 22.

$$\ell_{C1-O2} = C_1 - O_3 = 1{,}282\,\overset{0}{A};\ \ell_{(C-H)av.} \approx 1{,}110\,\overset{0}{A};\ \angle OCO = 109^{0};$$

$$\angle CCO = 125^{0}30';\ \angle CCH = 109^{0}28';\ \angle HCH = 109^{0}28'.$$

Considered radical is very interesting upon its electron structure. Both oxygen of the radical are equivalent both on the geometric parameters and on electron structure.

Like type of radicals have been discovered in experimental way only in a case of the γ–illumination of maleic acid at 77 ^{0}K [52]. On top of the occupied *MO* there are electrons of the unshared pairs, which occupy the P_y and P_z *AO* of oxygen atoms. General electron energy of the radical $CH_3C\overset{/\!/O}{\underset{\diagdown O^*}{}}$ is equal to

$-50{,}9281$ *a. u.*, the dipole moment $-3{,}86D$ and the occupations of the oxygen atoms are $1{,}8129$, $P_x = 1{,}1309$, $P_y = 1{,}6730$ and $P_z = 1{,}6905$.

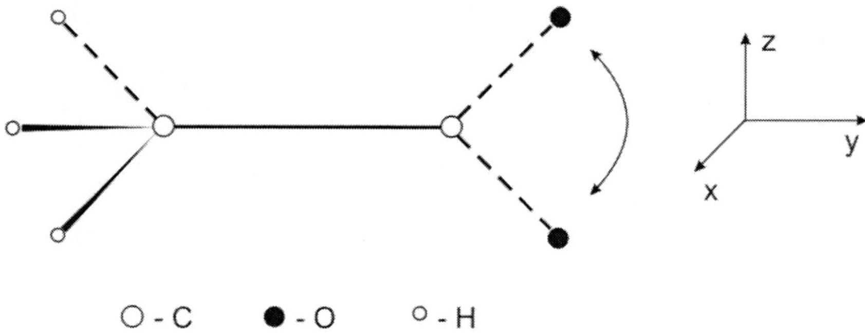

Figure 22. The optimal geometry of radical $CH_3C\overset{/\!/O}{\underset{\diagdown O^*}{}}$ $\ell_{C1-C4} = 1{,}456\,\overset{0}{A};$

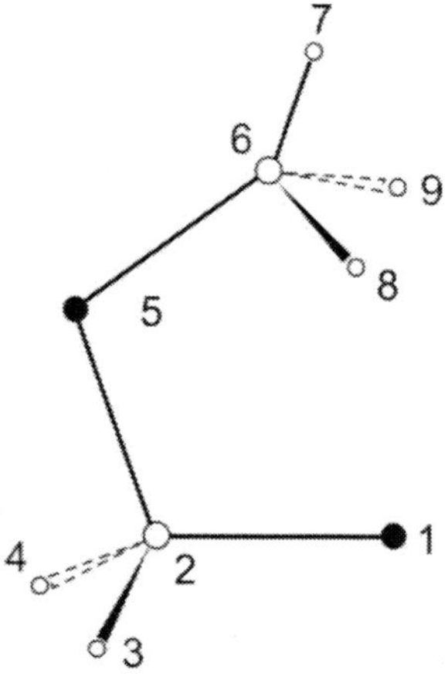

Figure 23. The optimal geometry of radical $H_2C\diagup^{OCH_3}_{\diagdown O^*}$.

2.6.4. Structure of Radicals $H_2C(OCH_3)O^{\cdot}$ and $(CH_3)_2NCH_2O^{\cdot}$

Optimal geometric parameters of radicals are represented on Figure 23 and Figure 24.

$$\ell_{(O1-C2)} = 1{,}3617\,\overset{0}{A}; \; \ell_{(C2-O5)} = 1{,}3805\,\overset{0}{A}; \; \ell_{(O5-C6)} = 1{,}3685\,\overset{0}{A};$$

$$\ell_{(C6-H7)} = \left(C_6 - H_8\right) = \left(C_6 - H_9\right) = 1{,}0962\,\overset{0}{A};$$

$$\ell_{(C2-H3)} = \left(C_2 - H_4\right) = 1{,}1200\,\overset{0}{A}; \; \angle O_1C_2O_5 = 106^0 51';$$

$$\angle C_2O_5C_6 = 108^0; \; \angle O_5C_6H_7 = O_5C_6H_8 = O_5C_6H_9 = 110^0 41';$$

$$\angle O_1C_2H_3 = O_1C_2H_4 = 110^0 46'; \; \angle H_3C_2H_4 = 109^0 28';$$

$$\angle H_3C_2O_5 = H_4C_2O_5 = 109^0 28'.$$

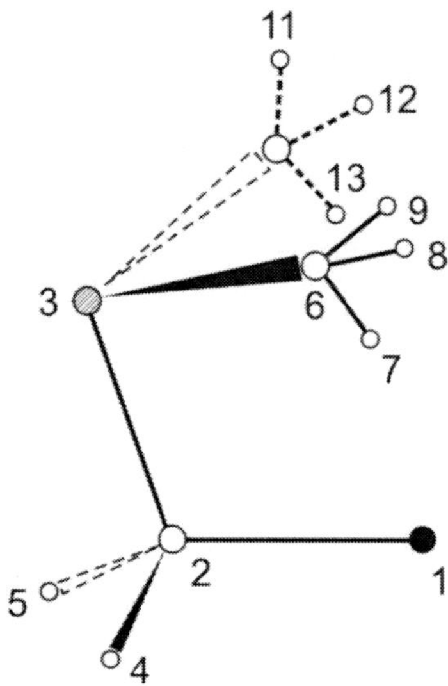

Figure 24. Optimal geometry of radicals $(CH_3)_2NCH_2O$. $\ell_{(O1-C2)} = 1{,}3600 \overset{0}{A}$;

$\ell_{(C2-H3)} = 1{,}1268 \overset{0}{A}$; $\ell_{(C2-H3)} = 1{,}4338 \overset{0}{A}$;

$\ell_{(N3-C6)} = (N_3 - C_{10}) = 1{,}4081 \overset{0}{A}$; $\ell(C-H)_{(CH3)} \approx 1{,}12 \overset{0}{A}$;

$\angle O_1C_2N_3 = 99^0 12'$; $\angle O_1C_2H_5 = 120^0 47'$; $\angle H_4C_2H_5 = 102^0 49'$;

$\angle H_5C_2N_3 = 114^0 5'$; $\angle C_2N_3C_6 = 114^0 58' = C_2N_3C_{10}$;

$\angle N_3C_6H_7 = 112^0 18'$; $\angle H_7C_6H_8 = 106^0 35'$; $\angle C_6N_3C_{10} = 116^0 36'$.

The less energetically advantageous is the such structure of radical $H_2(H_3C^4O^3)C^2O^1$, when the bonds O_3-C_4 and C_2-O_1 are located in perpendicular planes. *Trans*–disposition of vectors for bonds O_3-C_4 and C_2-O_1 is more advantageous, than the *cis*– one on 1,7 *kJ*, and in comparison with the above–pointed location on 3,8 *kJ*.

In the most advantageous structure *(trans*–form*)* of radical the free electron is located on P_x orbital, and *p*–unshared pair of electrons of final oxygen atom on P_y orbital.

Into nitrogen–containing radical a free electron is located on P_y of atomic orbital, and *p*–unshared pair on P_z *AO* of the oxygen atom.

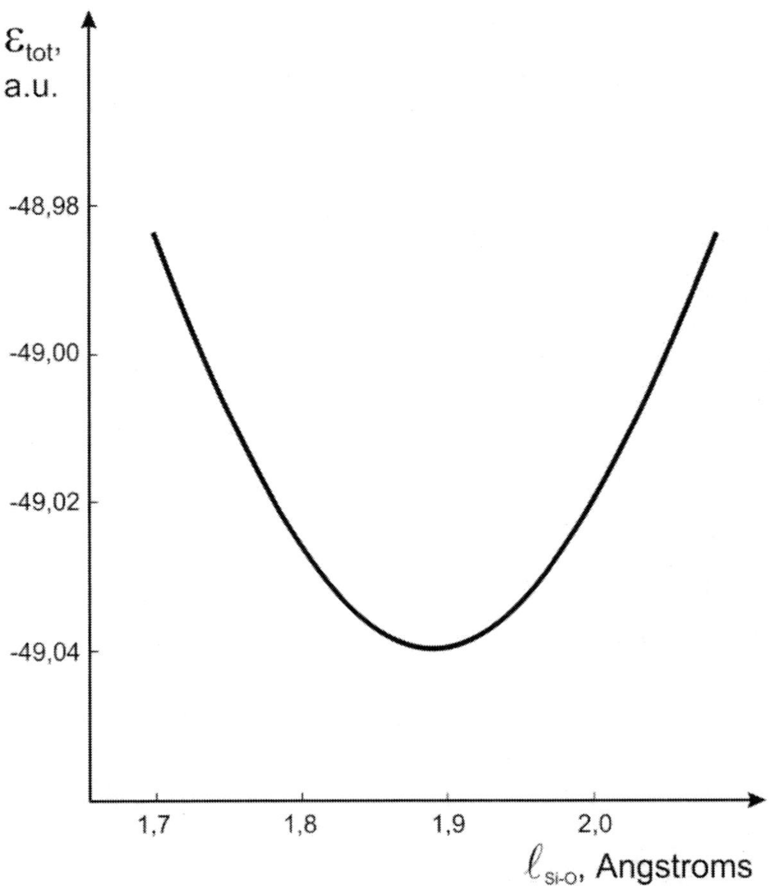

Figure 25. The dependence of full electron energy (E_{tot}) on length of bond $Si\text{–}O$ in trimethylsiloxyl radical.

2.6.5. Structure of Radical $(CH_3)_3SiO^{\bullet}$

It is follows from Figure 25, that the optimal distance of $Si\text{–}O$ in radical is equal to $1,9\,\overset{0}{A}$. Unpaired electron in this radical is enough strongly localized on MO with the energy $-0,6090$ $a.\,u.$

2.7. GENERAL ELECTRON CHARACTERISTIC OF FREE RADICALS

2.7.1. Dipole Moments

Generally it is known, that the dipole moments of the chemical particles depend on the spatial location of their different groups, the charges distribution on atoms and the interatomic distances.

It can be seen from the data of Table 10, that the values of the dipole moments for the oxyradicals *(RO)*, where *R* is the alkyl substitute, are some more than the experimental values of dipoles for the corresponding alcohols. This difference most likely is connected with the geometric structure of such compounds. For example, for CH_3OH the length of bond $C{-}O$ is equal to $1{,}428 \overset{0}{A}1$, and for radical CH_3O $\ell_{C-O} = 1{,}36 \overset{0}{A}$. For CH_3OH $\ell_{C-H} = 1{,}095 \overset{0}{A}$, and for CH_3O $\ell_{C-H} = 1{,}12 \overset{0}{A}$. The charges distribution on the atoms CH_3OH and CH_3O is also different.

Table 10. Values of the dipole moment for the alkoxyradicals and corresponding to them alcohols and acids

№	Radical RO	μ RO, D	Alcohol	μ of alcohol D [53]	Radical $RC{\overset{\displaystyle =O}{\diagdown}}_{O^{\bullet}}$	μ ROO, D	Acid	μ of acid D [53]
1	CH_3	1,73	CH_3OH	1,69	CH_3	3,86	$CH_3C{\overset{=O}{-}OH}$	1,73
2	C_2H_5	1,94	C_2H_5OH	1,70	C_2H_5	3,91	$C_2H_5C{\overset{=O}{-}OH}$	1,74
3	$n{-}C_3H_7$	1,98	$n{-}C_3H_7OH$	1,64	$n{-}C_3H_7$	4,45	$n{-}C_3H_7C{\overset{=O}{-}OH}$	1,9
4	$iso{-}C_3H_7$	2,05	$iso{-}C_3H_7OH$	1,58	$iso{-}C_3H_7$	–	$iso{-}C_3H_7C{\overset{=O}{-}OH}$	–
5	$n{-}C_4H_9$	2,01	$n{-}C_4H_9OH$	1,66	$n{-}C_4H_9$	4,34	$n{-}C_4H_9C{\overset{=O}{-}OH}$	–
6	$iso{-}C_4H_9$	–	$iso{-}C_4H_9OH$	1,63	C_6H_5	4,63	$C_6H_5C{\overset{=O}{-}OH}$	1,0
7	$tert{-}C_4H_9$	1,97	$tert{-}C_4H_9OH$	1,66	$CH_2{=}CH$	3,74	$CH_2{=}CHC{\overset{=O}{-}OH}$	–

However, in spite of the some difference in dipoles values for the compounds *ROH* and *RO*, it can be concluded, that the formation of radicals *RO* from *ROH* does not lead to the essential changes of σ–skeleton of the all molecule, whereas the σ–skeleton of radical $RC\diagup^{O}_{\diagdown O^*}$ is essentially differed from the σ–skeleton of the acids $RC\diagup^{O}_{\diagdown OH}$. It can be seen from the Table 10, that the values of dipole moments of radicals by *RC(O)O˙* type are differed more than in two times from those for the corresponding acids *RC(O)OH*.

2.7.2. Distribution of Charges into Radicals

The values of charges on oxygen atom having an unshared electron of radical *RO˙* should be depend on the nature of solvent *R*. At the donor properties of the substitutes increasing the values of the charges on oxygen are increased *(Table 8)*. It is follows from the data of Table 8 that the most advantageous conformation for radical ⬡–O˙ is the axial one, and for radical ⬡–O˙ the most energetically advantageous is the axial conformer *(«tub»)*.

Spin densities on the atom of oxygen *(Table 11)* for a series of radicals are changed within the range 0,94–0,97. Any others clear regularities of the *P* substitute nature influence on the spin density *(ρ)* are not observed.

Let us consider some apart the electron structure of the substituted alkoxyl radicals *(Table 12)* by $R_1CH(OR_2)O$ type.

Table 11. The values of charge (q), spin densities (ρ), dipole moments (μ) and full electron energies (E_{tot}) for alkoxyl radicals

№	RO$^\bullet$	q_c	$-q_O^\bullet$	$\rho_O^\bullet$	μ, D	$-E_{tot}$, a. u.
1	CH_3	0,2402	0,1775	0,9616	1,78	26,61055
2	C_2H_5	0,2520	0,1957	0,9672	1,94	35,0484
3	$n-C_3H_7$	0,2499	0,2011	0,9663	1,97	43,4852
4	$n-C_4H_9$	0,2469	0,2027	0,9647	2,01	51,9178
5	$iso-C_4H_9$	0,2452	0,2162	0,9706	2,07	51,9270
6	$tert-C_4H_9$	–	0,2267	0,9721	1,97	51,9331
7	 (axial)	–	0,2313	0,9073	1,72	57,1453
8	 (equat.)	–	0,2107	0,9462	2,36	57,1262
9	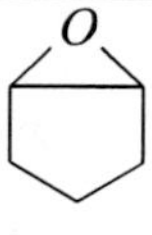 (equat. «tub»)	–	0,1987	0,9444	3,90	74,8257
10	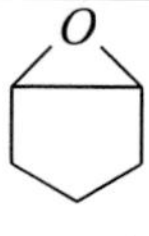 (equat. «seat»)	–	0,2085	0,9452	2,06	74,8214
11	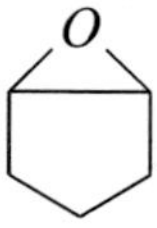 (axial. «tub»)	–	0,2357	0,9087	1,03	74,8476
12	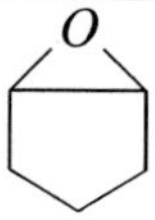 (axial. «seat»)	–	0,2151	0,9094	3,56	74,8396

As we can see from the data of Table 12, the diapason of the values of dipole moments for these radicals is the same as in a case of the usual alkoxyl radicals (Table 10). The values of spin densities on oxygen atom having the unpaired electron are within the interval 0,5–0,7. In a case, when $R_1 = $ (cyclohexyl ring), the radicals $R_1(R_2O^4)H^3C^2O^1$ can exist in two conformations, *namely* in axial and equatorial ones.

Table 12. The values of general electron energy (E_{tot}), dipole moment (μ), energy of *LFMO* and *HOMO* for radicals

№	Radical	$-E_{tot}$, a. u.	μ, D	$+\varepsilon_{HCMO}$, a. u.	$-\varepsilon_{B3MO}$, a. u.
1	cyclohexyl–CH(–OCH$_3$)–O• (axial.)	100,1307	1,60	0,4658	0,1216
2	cyclohexyl–CH(–OCH$_3$)–O• (equat.)	100,0903	1,94	0,4387	0,1313
3	cyclohexyl–CH(–OC$_2$H$_5$)–O• (axial.)	108,5682	1,50	0,4637	0,1245
4	cyclohexyl–CH(–OC$_2$H$_5$)–O• (equat.)	108,5277	1,86	0,4356	0,1348
5	cyclohexyl–CH(–OC$_3$H$_7$)–O•	117,0051	1,49	0,4627	0,1251
6	cyclohexenyl–CH(–OCH$_3$)–O•	96,8010	1,65	0,4696	0,1400
7	$н$-C$_6$H$_{13}$–CH(–OCH$_3$)–O•	103,2606	1,63	0,4705	0,1270
8	$н$-C$_6$H$_{13}$–CH(–OH)–O•	94,8185	1,67	0,4704	0,1283

Table 12. Continued

№	Radical	$-E_{tot}$, a. u.	μ, D	$+\varepsilon_{HCMO}$, a. u.	$-\varepsilon_{B3MO}$, a. u.
9	(axial.)	91,6883	1,64	0,4668	0,1227
10	(equat.)	91,6478	1,95	0,4397	0,1323
11	H_2C—$O\cdot$ with OCH_3	52,6492	1,69	0,5124	0,1069
12	H_2C—$O\cdot$ with OH	44,2042	–	0,5302	0,1062
13	H_2C—$O\cdot$ with $N(CH_3)_2$	55,4575	–	–	–

Comparing the series of the charges increasing on O_1 and O_4 (Table 13) in dependence on the substitutes nature it can be seen that they are not symbate. It is follows from this that the action of substitutes on the charges distribution in radicals is not clear inductive but by far complicated.

Table 13. Distribution of charges on a fragment $-C-\overset{O-C-}{\underset{H}{C}}-O\cdot$ **for radicals**

№	Радикал	$-q_{O1}$	q_{C2}	q_{C3}	$-q_{O4}$	q_{C5}
1	$O-CH_3$	0,2216	0,4124	0,0094	0,2989	0,2574
2	$O-C_2H_5$	0,2228	0,4105	0,0097	0,3081	0,2564

№	Радикал	$-qO1$	$qC2$	$qC3$	$-qO4$	$qC5$
3	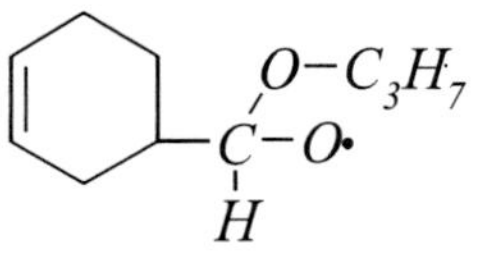 (циклогексенил)$-\overset{O-C_3H_7}{\underset{H}{C}}-O\bullet$	0,2235	0,4102	0,0099	0,3115	0,2543
4	$C_6H_5-\overset{O-CH_3}{\underset{H}{C}}-O\bullet$	0,2227	0,4074	0,0110	0,2970	0,2579
5	$н\text{-}C_6H_{13}-\overset{O-CH_3}{\underset{H}{C}}-O\bullet$	0,2221	0,4051	0,0081	0,3014	0,2571
6	$н\text{-}C_6H_{13}-\overset{OH}{\underset{H}{C}}-O\bullet$	0,2233	0,4072	0,0124	0,3395	0,1677
7	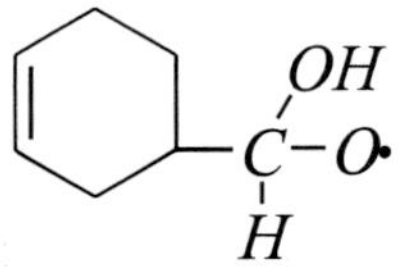 (циклогексенил)$-\overset{OH}{\underset{H}{C}}-O\bullet$	0,2229	0,4102	0,0132	0,3369	0,1683
8	$H_2C\overset{OOH}{-}O\bullet$	0,2426	0,4508	–	0,3082	0,2582
9	$H_2C\overset{OH}{-}O\bullet$	–	0,4248	–	–	0,3018

Table 14. The values of dipole moments _(μ)_,general electron energies _(E$_{tot}$)_, energies of _LFMO_, spin densities on oxygen atom for radicals

№	$R-C\underset{O}{\overset{O}{\diagdown}}$	μ, D	ρ	ε$_{LFMO}$, a. u.	−E$_{tot}$, a. u.
1	CH_3	3,86	0,5397	0,4229	50,9281
2	C_2H_5	3,91	0,5401	0,4167	59,3721
3	n–C_3H_7	4,45	0,4896	0,3625	67,7200
4	n–C_4H_9	4,34	0,4907	0,3656	76,1948
5	tert–C_4H_9	4,32	0,4829	0,3610	76,1880
6		3,82	–	0,4657	90,0136
7	C_6H_5	4,63	0,4897	0,3629	86,6159
8	$CH_2{=}CH$	3,74	0,5411	0,4189	57,5743
9	CH_2Cl	3,81	0,4969	0,3825	66,2975
10	CCl_3	3,75	0,5984	_	97,1575
11	Cl	3,55	0,66718	_	57,0013

Table 15. Polarografic characteristics of peroxides (I–X)

№	$-E_{1/2}$, B	$K \cdot 10^{-3}$	$\dfrac{E}{\lg I/I_n - I_{mv}}$	Dn
I	0,89	1,11	116	0,50
II	0,58	1,22	102	0,57
III	0,58	1,09	142	0,41
IV	0,59	2,33	135	0,43
V	0,57	2,03	123	0,43
VI	0,58	0,52	167	0,35
VII	0,49	2,72	171	0,34
VIII	0,46	1,12	188	0,31
IX	0,46	1,26	168	0,35
X	0,47	0,98	183	0,31

For radicals $R-C\overset{\displaystyle O}{\underset{\displaystyle O\cdot}{}}$ is typical that the atoms of oxygen in them are equivalent both on geometry and on electron structure *(see* Table 14*)*. The reinforcement of the donor properties of groups 1–5 leads to the some increasing of dipole moments of radicals and the acceptor properties of groups 9–11 leads to their decreasing.

As to the radical affinity to the electron, it is the most for the $R=C(CH_3)_3$ in a range of the compounds (1–5).

An increase of the donor properties in a range of the substitutes R_1 (compounds 1–5) leads to the increasing of the charges on oxygen atoms *(*Table 15*)*. Acceptor substitutes decrease of this charge (8–10). However, it is not explained from these positions the charge distribution in radical

. Most likely this is explained by the conjunction of the phenyl ring with the oxygen atoms.

2.8. Polarographic Investigation of some Peroxy Compounds

Presented Chapter has the purpose to explain the relationship between the electron structure of peroxides and their reduction potential, and also to obtain some information about the chemical mechanism of the studied peroxides reduction.

It is completely clear, that between the electron structure of organic compounds and their polarographic characteristics the definite relationship should be exist. In *ref.* [54] it was shown, that the reduction potentials of the organic compounds correlate with the energy of *LFMO*. Probably, like dependence should be expected also in a case of the peroxide compounds.

Swerns and *Zilberdt* [55] founded the definite qualitative dependence between the reduction potential and energy of bond of group *O–O* in peroxides. It was interesting to explain this dependence from the point of view of their electron structure.

Table 16. Geometric parameters of the optimal structure of equilibrium state of H_2O_2, its intermediate configurations and OH radical

№	ℓ_{O-O}, $\overset{0}{A}$	ℓ_{O-H}, $\overset{0}{A}$	$\angle\chi$, degree.	$\angle OOH$, degree.
1[*]	1,2223	1,040	90	108
2[**]	1,340	0,952	129	91
3	1,475	1,033	117	100
4	1,675	1,033	117	94
5	1,875	1,033	117	90,5
6	1,975	1,033	117	88
7	2,075	1,033	117	85
8	2,175	1,033	117	83
9	2,275	1,033	117	83
10	2,375	1,033	117	85
11	2,675	1,033	117	86,5
12	3,075	1,033	117	88
13	3,275	1,033	117	93
14[***]	–	0,971	–	–

[*] Equilibrium structure *(CNDO)*.

[**] Equilibrium structure *(FNDP)*.

[***] Experimental length of the bond of hydroxyl radical.

The investigations objects were: *tert.*–butylperoxytetrahydrobenzoate (*I*), *tert.*–butylperoxymethyl–3,4–epoxytetrahydrobenzoate (*II*), *tert.*–butylperoxymethyldi-methylamine (*III*), *tert.*–butylperoxymethyldiethylamine (*IV*), *tert.*–butylperoxy–N–piperydine (*V*), *tert.*–amylperoxymethyldimethyl-amine (*VI*), *tert.*–amylperoxy-methyldiisopropylamine (*VII*), *tert.*–butylperoxycyclopentene–2 (*VIII*), *tert.*–hexylperoxycyclopentene–2 (*IX*) and *tert.*–hexylperoxy–2,3–epoxycyclopentane (*X*). The investigations were carried out with the use of the mercury dropping electrode. It has been used The mercury capillary was used with the following characteristics: $m = 2,34$ mg/s; $\tau = 3,34$ s; m 2/3 τ 1/6 = 2,16. The saturated calomel electrode was the reference electrode.

The mixed character of the polarographic wave is confirmed by the linear dependence of the maximum current *(I_n)* from the root square of high of the mercury, not passing via the point of origin. Potentials of the semi–waves of reduction determined graphically from the equality $lgI/I_n–I = 0$ and from the polarograms are near per values *(*Table 16*)*.

Upon the variation of peroxides concentration the value of potentials $E_{1/2}$ are little changed. The electrons consumption per mole of the reducing peroxides *I–X*, and also the values of the angular coefficients of waves for peroxides *I–X* were founded from the equation of the polarographic wave (Table 15). Graphic dependence of $lgI/I_n–I$ on the potential of mercury dropping electrode (with respect to the saturated calomel one) has the straight character. Obtained fractional values for a number of electrons points on the irreversibility of the reduction process, realizing on the mercury drop [56].

As a result of the electroreduction of the *tert.–*butylperoxymethyldimethylamine it was discovered the *tert.–*butyl alcohol, tetramethylethylendiamine and water (one molecule of water per one molecule of peroxide), and also the traces of dimethylformamide. Starting from the data of the chromatographic analysis of the final products of electrolysis, the electroreduction scheme of peroxide *III* can be represented as follows:

$$(CH_3)_2NCHOOC(CH_3)_3 \xrightarrow{2H^+ +2e^-} HOC(CH_3)_3 + (CH_3)_2NCH_2OH$$

$$2\,(CH_3)_2NCH_2OH \xrightarrow{-2H_2O} \begin{matrix} (CH_3)_2NCH \\ \| \\ (CH_3)_2NCH \end{matrix}$$

$$(3)$$

At the peroxides electrolysis of *tert.–*butylperoxycyclopentene–2, *tert.–*butylperoxytetrahydrobenzoate and *tert.–*butylperoxytrimethylsilane the cyclopente-nol–2, tetrahydrobenzoic acid, trimethylsilanol and *tert.–*butanol are respectively formed in all cases. So, the electroreduction scheme of these listed peroxides, from our point of view, can be represented in following way:

$$ROOC(CH_3)_3 \xrightarrow{2H^+ +2e^-} ROH + HOC(CH_3)_3$$

$$R = \; \text{⬠} \; , \; \text{⬡}_{C\overset{O}{\diagdown}} \; , \; -Si(CH_3)_3 \; .$$

$$(4)$$

Based on data of the chromatographic analysis of the final products of methoxy*tert.–*butylperoxycyclohexenylmethane electrolysis, the electroreduction scheme is as follows:

$$\text{(structure)} - C(H)OOC(CH_3)_3 \xrightarrow{\ 2H^+ + 2e^-\ } HOC(CH_3)_3 + \text{(structure)} + H\overset{O}{\overset{\|}{C}}OCH_3 \tag{5}$$

Probably, two electrons take part at the polarographic reduction of the investigated peroxides.

From general considerations it should be expected some correlation between the reduction potential of semi–wave of peroxides and value of the energy of *LFMO*, *i. e.* the affinity to the electron with the reversed sign.

The relationship of the reduction potential $E_{1/2}$ of different peroxides with their activation energy of the decomposition reaction is illustrated by data of Table 17 [55].

It is follow from data of Table 17 that such relationship is merely qualitative. If to assume, that between the peroxides reduction potential and energy of *LFMO* should be exist some definite dependence, then, probably, necessary to expect the definite relationship between the energy of *LFMO* and energy of the activation process. However, here one condition should be fixed, connected with the location of energetic levels in peroxy compounds. The level of *LFMO* in different peroxides can correspond to different chemical bonds.

Table 17. The values of the reduction potentials and decomposition activation energies for some peroxides

№	Peroxide	$E_{1/2/}$	$E_{act.,}$ kJ/mole
1	Ditert.–butylperoxide	2	159–167
2	Dialkylperoxides	1	150–155
3	Tert.–butylperether	0,8–1,0	146–150
4	Hydroperoxides	0,6–0,9	113–134
5	Diacylperoxides	0,1	126

It is clear, that quite logically to expect the correlation between the energy of the loosening level of *O–O* bond in peroxides and the activation energy of process. If the loosening levels ε_{O-O} are by the *LFMO* for peroxides, then it can be said about the relationship between the reduction potential and these levels. It can be seen from the data of Table 18 that for the represented peroxides the symbate dependence between the $E_{1/2}$, ε_{O-O} and $E_{act.}$ is really observed.

Table 18. Dependence between the energy of loosening *MO*, which is responsible for the *O–O* bond, reduction potential and activation energy for some peroxides

№	Compound	$-E_{1/2}$	ε_{LFMO}, a. u.	$\varepsilon_{loosening.}$ O–O, a. u.	$E_{act.}$, kJ/mole	$E_{stab.}$, a. u.
1	$H_2\overset{\diagup OCH_3}{COOC(CH_3)_3}$	–	0,1822	0,1822	175,7	1,0508
2	$CH_3\overset{O}{C}OOC(CH_3)_3$	1,02	0,1592	0,1727	159,0	1,0056
3	$C_6H_5\overset{O}{C}OOC(CH_3)_3$	0,95	0,1184	0,1720	150,6	0,9658
4	$C_2H_5\overset{O}{C}OOC(CH_3)_3$	0,90	0,1606	0,1710	146,4	0,9554
5	$(CH_3)_2NCH_2OOC(CH_3)_3$	0,58	0,1696	0,1636	133,9	0,9203
6	$CH_3\overset{O}{C}OO\overset{O}{C}CH_3$	0,1	0,1461	0,1477	125,5	0,8900

However, such dependence is quantitatively difficult to imagine, since in peroxides there are some loosening levels, which are responsible for the *O–O* bonds. In Table 18 there is one among the most probable ones, mainly respondent for the *O–O* bond.

Observed symbate dependence (Table 18) in principle has the base, but for many cases it can be broken in connection with the *MO* delocalization degree for the loosening level $\varepsilon_{O–O}$. If the *MO* delocalization degree of the loosening level $\varepsilon_{O–O}$ is approximately the same in a series of the peroxides, than the dependence between $\varepsilon_{O–O}$, $E_{1/2}$ and $E_{act.}$ should be performed in the best way.

Most likely, such regularities of the relationship between the energy of *LFMO* and homolysis activation energies are partial and are performed only for the determined series of peroxides having the similar structure of the reactive center. Peroxides, represented in Table 18, are conformational similar, so the structure of their reactive center can be represented in *cis*–conformation $R_1R_2\,C\overset{\diagup X\cdots O_\beta-R}{=O_\alpha}$. As it will be shown later (Chapter 3.11), the activation energy of the peroxides homolysis is determined by not only the strength of the $O{-}O$ bond, but also by the interaction of atoms of the reactive center valence free between themselves *(C...O$_\beta$, X...O$_\beta$, X...Oα)*.

Orbital energy or so–called stabilization energy accepted as the energetic total criterion taking into account the interactions of atoms in $O{-}O$ bond and interaction of valence free atoms of the reactive center in peroxides. The direct dependence is observed between this characteristic and the activation energy of homolysis for a series of peroxides. Therefore, between the reduction potential of peroxides with conformational similar structure of the reactive center and the stabilization energy the straight dependence should be performed *(*Table 18*)*.

Summarizing the above–said as to the structure of considered organic peroxides and oxy–radicals, it can be done some general conclusions useful for the solution of the main question concerning to relationship between their structure and reactivity.

An existence of the cyclic–containing peroxides (cycloperoxyacetals, cyclopentenyl peroxides) in *bi–* conformations (axial and equatorial) beforehand foresees some peculiarities of homolysis kinetics of these forms related, first of all, with the different durability of the peroxy bond.

Conformational stability of the axial conformers for peroxyacetals is explained by the spatial interaction of the α–atom of oxygen of peroxy bridge and double bond of the cyclohexene ring. With the use of the conformational analysis methods and also using the calculated and experimental *IR–* spectroscopy it was proved that for the peroxides by general formula

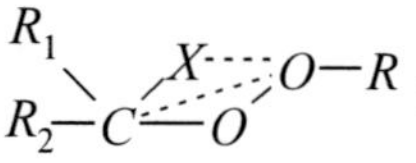

$$\begin{array}{c} R_1 \\ \diagdown \\ R_2{-}C \end{array}\!\!\!\!\!\!\!\! \begin{array}{c} X\cdots\cdots O{-}R \\ \diagup\quad\diagup \\ {=}\!{=}O \end{array}$$ more energetically advantageous is the *cis*–form of bond *C–*

X with respect to the peroxy one. The stability of *cis*–conformations is caused by the interaction of valence free atoms of reactive center *(X...O$_\beta$, C...O$_\beta$, X...Oα)*. The interaction energy of valence free atoms it is proposed to be characterized by the orbital energy, which is a measure of the peroxide reactive center stabilization at the expense of the charge transfer effects between valence free atoms.

It is quite logically to expect the relationship between this characteristic and activation energy of the peroxides homolysis and their reduction potential.

Consideration of electron structure of oxy–radicals by quantum–chemical method leads to the conclusion that their electron characteristics (the charges on the atoms, spin density and *ect.*) depend on the interactions of valence free atoms, *i. e.* conformations.

Chapter 3

REACTIVITY OF PEROXY COMPOUNDS IN HOMOLYSIS REACTIONS

Proposed in a previous Chapter aspect of the organic peroxides structure enough clear states a set of the concrete problems in a field of their reactivity.

The questions of the peroxides reactivity in homolysis reactions (a stabilization of the activated complex, a determination of the reactivity indexes, the influence of the substitutes nature, the estimation of pre–exponential factors *and ect.*) requires an information about the activated complex. Single source for obtaining of this information is the theoretical calculation of the reaction way. Under the aspect of the theoretical studies of ways of peroxy decomposition on free radicals there are only separate works [51]. The last work was appeared in printing in 1977. At the same time it were printed also our works [57–60].

Authors [61–64] carried out the quantum–chemical calculation for a set of the organic peroxides into equilibrium state. However, the role of the conformation via their homolysis process is not considered. The consideration of homolysis of the simplest peroxy compounds in theoretical aspect can be by some model for more complicated peroxy systems.

3.1. RADICAL DECOMPOSITION OF HYDROGEN PEROXIDE AND DIMETHYLPEROXIDE

For determination of the optimal structure of H_2O_2 it has been used the experimental data on its geometry

$$\left(\ell_{O-O} = 1{,}475\,\overset{0}{A}, \ell_{O-H} = 0{,}96\,\overset{0}{A}, \angle OOH = 94^0, \chi = 111^0 \right)$$ [65].   At   the

determination of the optimal structure of the intermediate systems of H_2O_2 via reaction the conditions of equality of lengths O–H of bonds and the angles' OOH were imposed. The length of bond –O–O– was varied within the limits of 0,875–3,275 $\overset{0}{A}$. In Table 16 there are data on geometry of the equilibrium state of H_2O_2 and its intermediate complexes *(IC)*. Only starting from the length of bond –O–O– in *IC* equal to 1,675 $\overset{0}{A}$ the length of bond O–H is practically near to the length of the bond in radical OH. At the lengthening of the bond –O–O– the value of angle OOH passes via the minimum *(see* Table 16)*.* The value χ is changed only during the start period of time.

At the calculation of optimal structure for CH_3O and CH_3OOH_3

$$\ell_{O-O} = 1{,}49\,\overset{0}{A}, \qquad \ell_{C-O} = 1{,}41\,\overset{0}{A}, \qquad \ell_{C-H} = 1{,}095\,\overset{0}{A}, \qquad \angle COO = 100^0,$$

$\angle HCH = 109^0 28'$ were taken as the null approximation. Optimal structure of CH_3O is satisfactory agreed with data of *ref.* [66]. During the calculation process of *IC* for CH_3OOCH_3 the relationship was changed in a ranges of 0,83–3,4 $\overset{0}{A}$. *IC* has been optimized in a range of 1,236–3,4 $\overset{0}{A}$.

General electron energy *(E_{tot})* for equilibrium state H_2O_2 and OH is respectively equal to −36,7508 and −18,162 *a. u.* For CH_3OOCH_3 and CH_3O E_{tot} is equal to −53,5890 and −26,0987 *a. u.* The dissociation energy of H_2O_2 is equal to −0,4257 *a. u. (1112,9 kJ).* For CH_3OOCH_3 this magnitude is equal to 0,3816 *a. u. (1000,0 kJ).*

If to consider for the gaseous phase the dissociation energy approximately equal to the activation energy of the dissociation process, then it can be seen the evident discrepancy for the theoretically calculated dissociation energies for H_2O_2 and CH_3OOCH_3 with the experimental values −200,8 и 154,8 *kJ/mole* respectively [67].

On Figure 26 there are profiles of the optimal surfaces of energy along the homolysis reaction way for H_2O_2 and CH_3OOCH_3. Starting only from the analysis of the profiles of potential surface, it is presented no possible to estimate the geometry of the critical intermediate configuration *(CIC).* However, it can be assumed, that the *CIC* is located in a field of these curves inflection $\left(1{,}8 - 2{,}2\,\overset{0}{A} \right)$.

Let us consider the general regularities of H_2O_2 and CH_3OOCH_3 homolysis. On Figure 27 there are changes of the charge values, occupancies of $s-$ and $p-$orbitals of oxygen atoms on the coordinate of reaction.

For the all curves into proposed field of the *CIC* existence it is observed the inflection. In a field of $2\overset{0}{A}$ it is observed the *HEMO* and *LFMO* energy change corresponding to the bond $-O\!-\!O-$, and the energy of the remaining *MO* is little changed [59]. This once more says about fact that in this field there is *CIC*. At the internuclear distance of $-O\!-\!O-$ equal to $3,075\overset{0}{A}$ the energy of intermediate complex *HO...OH* (–36,3252 *a. u.*) and of two *OH* radicals is the identical, *i. e.* under presented conditions there are formed exactly free radicals *OH*.

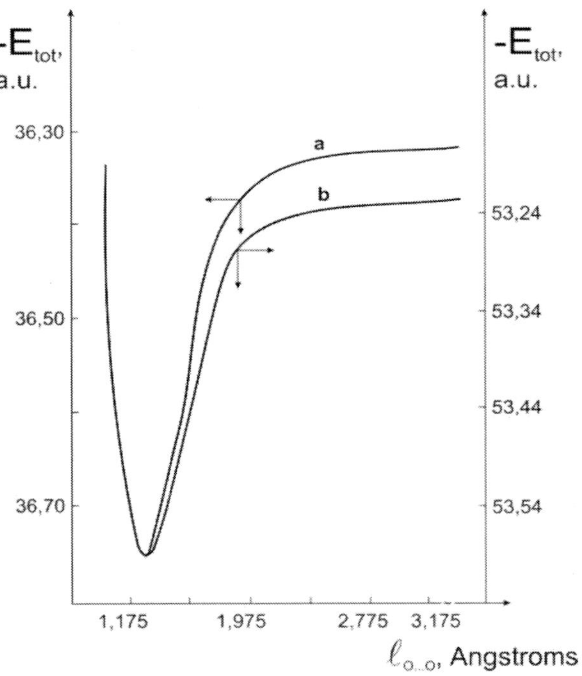

Figure 26. The change of general electron energy of molecules H_2O_2 (curve *a*) and CH_3COOCH_3 (curve *b*) along the coordinate of reaction.

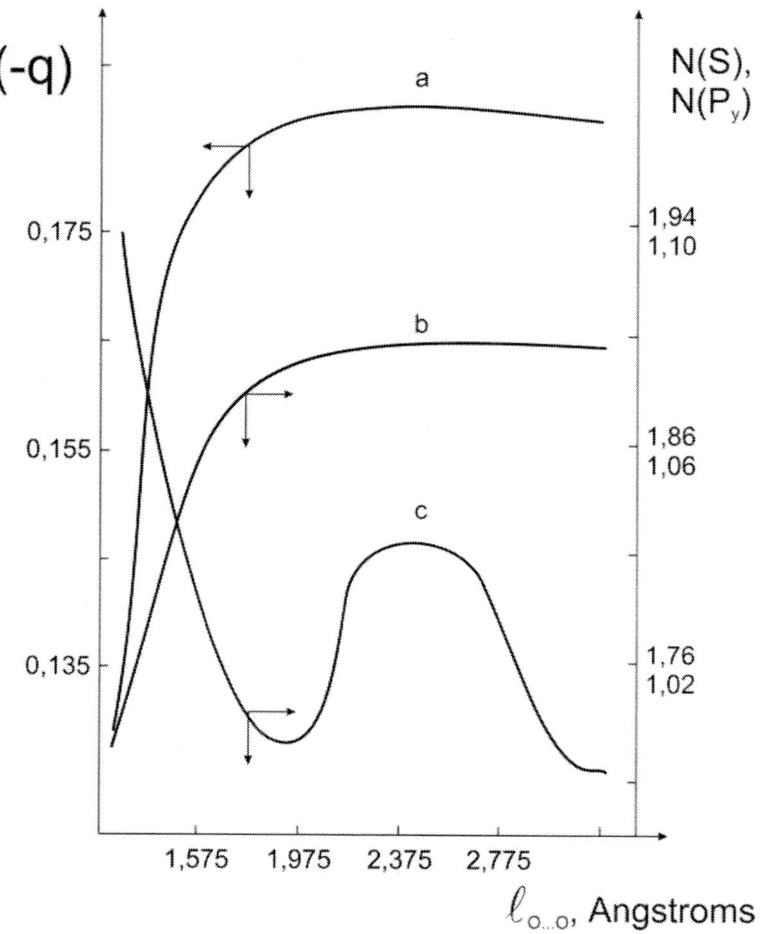

Figure 27. The dependences of: a – charge, b – occupancy of s–orbitals $N(S)$ and c – occupancy of p_y–orbitals $N(P_y)$ of oxygen atoms in H_2O_2 on the distance between themselves.

At $\ell_{O...O} \rightarrow \infty$ the energy is the same. On the basis of curve presented on Figure 27 it can be concluded that it is formed the enough polar structure.

The same decomposition regularities are observed also for CH_3OOCH_3 and are represented on Figure 28. The practical coincidence of the energies values of MO for IC of CH_3OOCH_3 and for free radical is observed in a field of 2,6–3,0 $\overset{0}{A}$ [59].

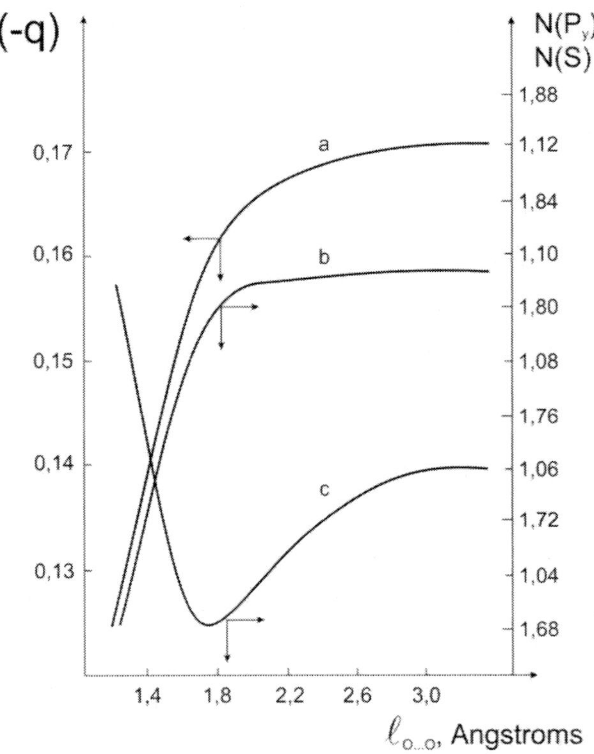

Figure 28. The dependencies of: a – charge, b – occupancy of s–orbitals $N(S)$ and c – occupancy of p_y–orbitals $N(P_y)$ of oxygen atoms in CH_3OOCH_3 on the distance between themselves.

3.2. NON–EMPIRICAL DESIGNING OF THE HYDROGEN PEROXIDE HOMOLYSIS

As it was shown into previous Chapter, the semi–empirical methods of the calculations do not permit with the sufficient exactness to determine the activation energy even for the simplest peroxides H_2O_2 and CH_3OOCH_3. In *ref.* [60] we have done the attempt to calculate of this value by *ab initio* calculation for the molecule of H_2O_2.

The calculations were carried out with the use of the *FUGA* programme [68] in compressed *Gaussian* basic set $2(S)2(P)$ on the atoms of oxygen and $1S$ on the atoms of hydrogen. The geometry was taken experimental. In Table 19 it is presented the dependence of the full electron energy on the internuclear distance of $–O–O–$.

Table 19. Dependence of the full energy of H_2O_2 (E_{tot}, a. u.)

on distance $\ell_{O-O}\left(\overset{0}{A}\right)$

ℓ_{O-O}	1,425	1,475[*]	1,525	1,675	1,825	1,875
E_{tot}	−150,3114	−150,3137	−150,3114	−150,2866	−150,2474	−150,2331

[*] Experimental value of the length of bond −O−O− in H_2O_2.

The value of $\ell_{O-O} = 1,875\,\overset{0}{A}$ is the maximal at which the calculation agreement procedure for the system with the closed shell is converged. The following increasing of the −O−O− distance leads to the divergence of the self–agreement procedure of the system.

An energy of the OH radical, starting from the energy of $\overset{+}{O}H$, with the use of the *Kupmanns's* theorem [69] is equal to −75,1166 a. u., and its double value −150,2332 a. u. is in good agreement with the limited value of energy of H_2O_2 at $\ell_{O-O} = 1,875\,\overset{0}{A}$. These calculations once more confirm that at $\ell_{O-O} = 1,875\,\overset{0}{A}$ we have deal with the radical system.

The activation energy of the H_2O_2 dissociation process under this approximation is equal to $(-150,2331) - (-150,3137) = 0,0806$ *a. u. (209 kJ)*, i. e. it is observed a good agreement with the experimentally founded value 201 *kJ/mole*.

Summarizing the above–said, it can be concluded that the *ab initio* calculations permitted with the sufficient exactness to estimate the activation energy of the peroxides homolytic decomposition reaction. However, the complications of the calculations scarcely made of this method quite accessible in the nearest possible time for more complicated systems.

Based on the carried out calculations it can be assumed that the *CIC* is formed at the internuclear distance −O−O− in H_2O_2 $1,8 \pm 0,1\,\overset{0}{A}$.

3.3. DESIGNING OF THE PROCESS OF RADICAL DECOMPOSITION FOR THE MOLECULE OF METHYLPERACETATE

Investigation of the H_2O_2 and CH_3OOCH_3 homolysis in theoretical aspect can to some extent design the reactions of more complicated peroxy systems by dialkyl type only. However, there are other kinds of the peroxy compounds, *for example*, peresters $R-\overset{O}{\underset{}{C}}-OOR_1$ and diacyl peroxides $R-\overset{O}{\underset{}{C}}-OO-\overset{O}{\underset{}{C}}-R_1$, reactions of which due to their structure specification hardly can be designed by the reactions of the simplest H_2O_2 and CH_3OOCH_3. It seems to us, that the above–mentioned peroxy systems can be designed by the enough simple system of methylperacetate $H_3C-\overset{O}{\underset{}{C}}-OOCH_3$.

The optimal elements of the structure in equilibrium state of the methylperacetate are represented in Table 20.

On Figure 29 there is dependence of the full electron energy of methylperacetate on the internuclear distance of $-O-O-$ *(curve I)*.

Via the $H_3C-\overset{O}{\underset{}{C}}-OOCH_3$ decomposition reaction the lengthening of the $C–C$ and $C=O$ bonds and the shortening of the $O–CH_3$ bond proceed. In radical $H_3C-\overset{O}{\underset{}{C}}-O\cdot$ the bonds $C–O$ are equal. The change of the $CH_3(CO)–O$ bond proceeds only to the internuclear distance of $-O-O-$ 1,85 $\overset{0}{A}$. At more $\ell_{O...O}$ it equal to 1,36 $\overset{0}{A}$ and corresponds to the length of a bond into oxyradicals. It can be seen from Figure 29 that at the internuclear distance 1,8 $\overset{0}{A}$ under methylperacetate homolysis it is observed the maximum of the orbital energy and the minimum of the electrons repulsion energy. This means, that in *CIC* the critical distance is equal to 1,8 $\overset{0}{A}$.

Table 20. Geometric parameters of the methylperacetate molecule into equilibrium and activated states

№	Bond	The length of bond $\overset{0}{A}$			Valence angle, degrees.	The value of angle, degree		
		Zero approxim.	Optimal values			Zero approxim.	Optimal values	
			equilibrium state	activated state			equilibrium state	activated state
1	O–O	1,49	1,246	1,850	OOC	100	101,3	99,5
2	O–C	1,44	1,388	1,348	OOC	100	101,3	100
3	O–C	1,41	1,378	1,360	(O)CO	125	115	117
4	C=O	1,24	1,270	1,278	OCC	125	125	119
5	C–O	1,52	1,441	1,452	(O)CC	110	120	124
6	$(C–H)_{cp}$	1,09	1,122	1,122	$(OCH)_{cp}$ $(CCC)_{cp}$ $(HCH)_{cp}$	109,5	109,5	109,5

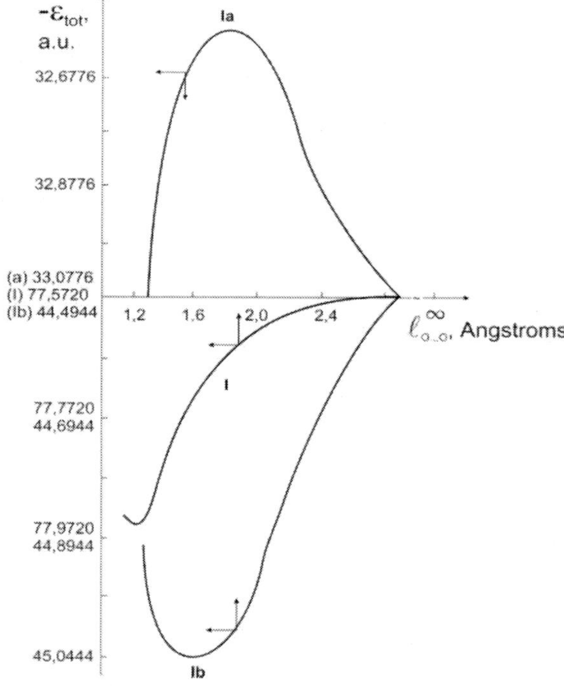

Figure 29. Change of the full electron energy (I), orbital energy *(Ia)* and the energy of the electrons repulsion *(Iб)* for $H_3C\text{-}\overset{\displaystyle O}{\overset{\|}{C}}\text{-}OOCH_3$ on the reaction way.

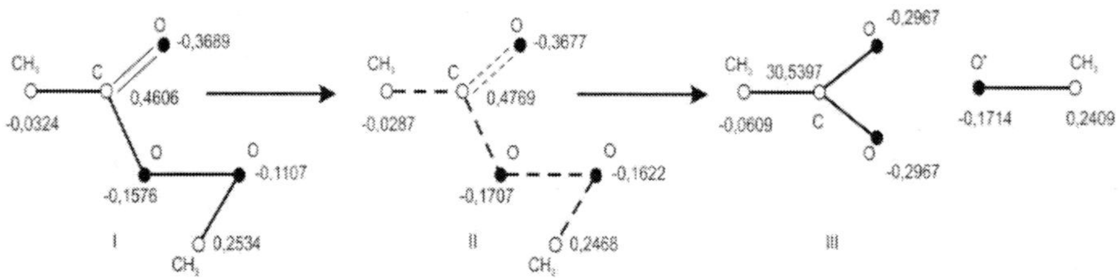

Figure 30. Distribution of charges in methylperacetate in equilibrium state (I), activated complex (II) and final products (III).

At the internuclear distance $\ell_{O-O} = 3\overset{0}{A}$ the sum of the energies for the corresponding radicals correlates with the energy of *IC*; this means that the free radicals are formed under this distance.

So, summarizing the above–said it can be concluded that, *firstly*, *CIC* for more complicated systems is formed also in a field of $2\overset{0}{A}$; *secondly*, the charges distributions for peresters via homolysis reactions are differed from the equilibrium structure and the final radical products upon their electron structure *(see* Figure 30*)*.

3.4. ABOUT POTENTIAL ENERGY OF H₂O₂ DECOMPOSITION WITH TAKEN INTO ACCOUNT THE ROTATION ENERGY

As it can be seen from the previous Chapters, the geometry of activated complex cannot be exactly determined starting from the surface of the potential energy calculated by semi–empirical quantum–chemical method. However, if to the value of the potential energy to add the molecules rotation energy for one among quantum numbers and for every distance between $-O-O-$, then we will obtained the curve with the maximum *(see* Figure 31*)*, which corresponds to the activated complex.

The rotation energy of H_2O_2 on the reaction way was calculated as for two–atomic molecule. General energy *(ε)* of the molecule H_2O_2 on the reaction way is as follows:

$$\varepsilon = E_{tot} + \frac{h^2}{8\pi^2 \mu r^2} j(j+1) \tag{5}$$

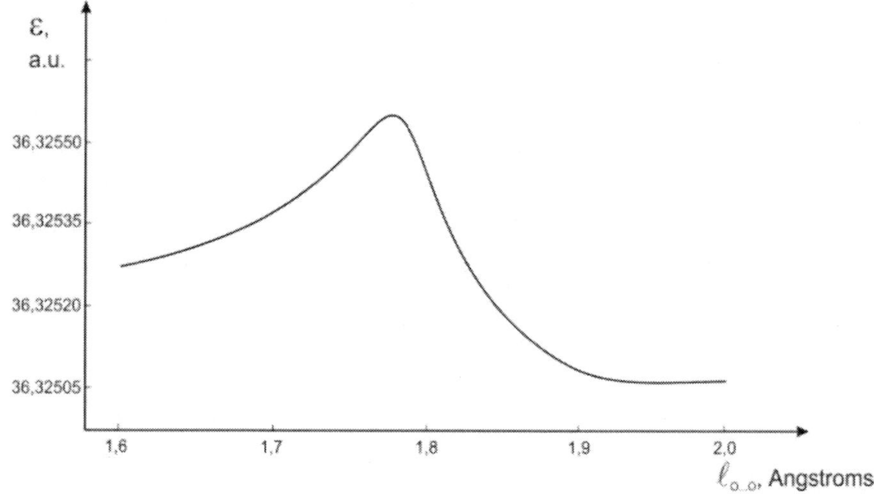

Figure 31. A profile of the potential surface of H_2O_2 homolytical decomposition with taken into account the rotation energy.

From the condition of maximum on the curve *(Figure 31), i. e.* $\dfrac{d\varepsilon}{dr} = 0$ in accordance with the equation (5) it is determined the interatomic distance, corresponding to the indicated maximum for the presented j. Repeating the calculations for different r, we will obtain the dependence of j on r_j *(Figure 32)*.

The most probable value was determined from the distribution of probability $(2j+1)e^{-E_j^*/RT}$ [70], where $(2j+1)$ is the multiplicity of state and E_j^* is the maximal value of the rotation energy with respect to the energy of the separate atoms. On Figure 33 it is seen that $E_j^* = E_j + E_{tot} + (-D)$. The dissociation energy was taken equal to 11,4 e. v, and the temperature 873 K. The dependence of E_j^* on j is represented on Figure 34. The most probable quantum number j corresponds to the maximum on this curve.

So, the activated complex at the radical decomposition of H_2O_2 is formed under the internuclear distance $1,8 \overset{0}{A}$.

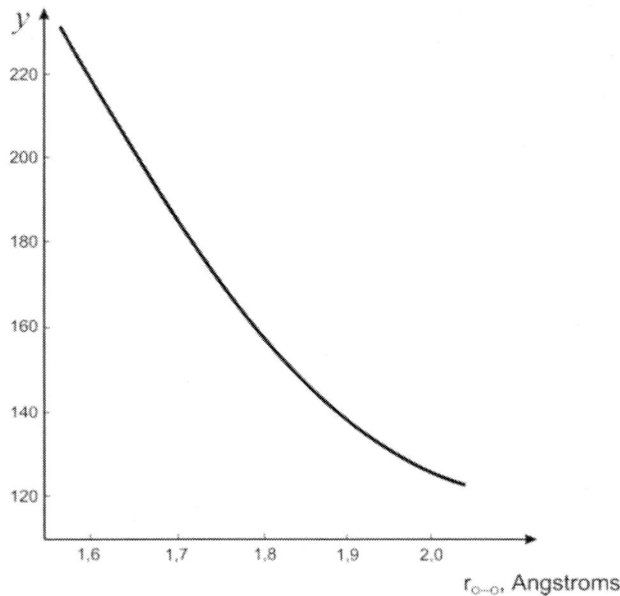

Figure 32. The dependence of the rotation quantum number value j on the internuclear distance $-O-O-$ into molecule of H_2O_2.

Considered example has the illustrative character. The calculation of more complicated molecules under the like aspect is conjugated with a series of well–known complications.

Probably, the fact of the maximal rapprochement of the connecting and antibonding MO, corresponding to the bond $-O-O-$ in H_2O_2 at the internuclear distance $-O-O-$ $1,8 \overset{0}{A}$, is evidence of the activated complex formation in this field [71].

At last, it is necessary to note, that about the lengthening of bond into the activated complex it can be predetermined based on its volume $\Delta V^{\neq}$ increasing. For example, under decomposition of the benzoil peroxide $\Delta V^{\neq} = 10 sm^3 / mole$, that corresponds to the $-O-O-$ bond distance lengthening into the activated complex on $0,4 \overset{0}{A}$ [72]. Our calculations are satisfactory agreed with this result. Author of $ref.$ [51] has been done the same conclusion under calculations of the methylhydroperoxide decomposition. However, the potential curve and the author's considerations are not presented in this work.

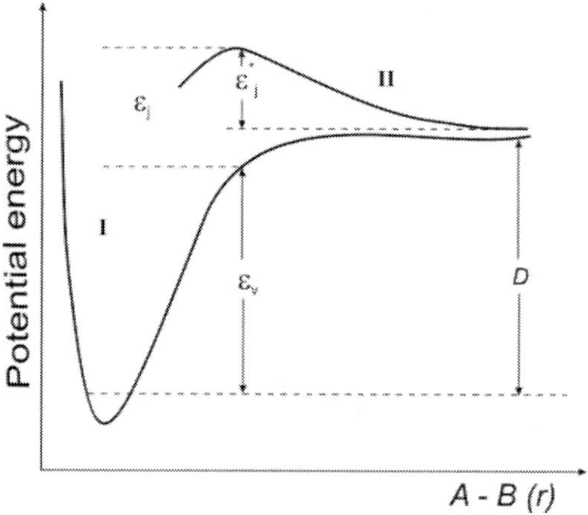

Figure 33. Potential energy (I) of the two–atomic molecule; potential energy (II) with taken into account the rotation energy.

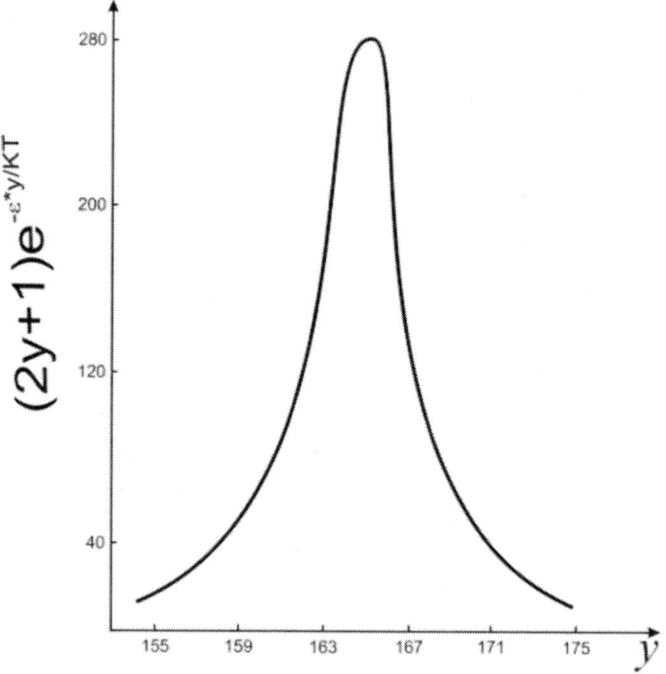

Figure 34. The curve of the most probable distribution of the rotation quantum number into the activated state $\left(\ell_{O-O} = 1{,}775 \overset{0}{A} \right)$.

3.5. DESIGNING OF THE TERT.– BUTYLPEROXYMETHYLDIMETHYLAMINE MOLECULE HOMOLYSIS

Above it were considered the general peculiarities of the homolytical decomposition of not great molecules of peroxides. It was interesting to trace the change of the quantum–chemical characteristics on the way of more complicated unsymmetrical peroxide decomposition. The nitrogen–containing peroxide $(CH_3)_2NCH_2OOC(CH_3)_3$ has been used as such object of the investigation. It can be seen from the Table 21 that the atom of nitrogen greatly influences on the distribution of electron density of the peroxide bridge, *i. e.* already into equilibrium state it is observed some distribution of the negative charges. With the internuclear distance of $O–O$ increasing the charges on the oxygen atoms are increased as a result of the localization of electrons. However, the increase of charge on the oxygen atom O_6 is more intensive in connection with more decoupling of fragment $...\underset{6}{O}...\underset{7}{C}(CH_3)_3$

for the transition state $(CH_3)_2N_1\cdot\cdot\cdot{}_2\overset{\overset{\textstyle H_3}{\textstyle |}}{\underset{\underset{\textstyle H_4}{\textstyle |}}{C}}\cdot\cdot\cdot\underset{5}{O}\cdot\cdot\cdot\underset{6}{O}\cdot\cdot\cdot\underset{7}{C}(CH_3)_3$ and its fitting to the

free radical $OC(CH_3)_3$. During the reaction process the charges on C_2 and C_7 are some decreased as a result of the partial delocalization of electron density upon corresponding $C–O$ bonds.

Let us note, that the decrease of the charge on the atom of nitrogen in *KPK* and further on the reaction way testify to the «bond» of the nitrogen atom with the O_6- oxygen of peroxy bridge is some weakened in these states.

It can be seen from the data of Table 21 that under assumption of the synchronous break of the bonds the great changes into distribution of the charges proceed only into nitrogen–containing part of the peroxide molecule.

Comparing the differences in full electron energies for transition and starting states of these two mechanisms, it can be concluded, that the synchronous mechanism is less energetically advantageous. Really, in the first case $E_1 = -107,3797 + 107,56258 = 0,1829$ *a. u.*, and in the second case $E_2 = -107,31744 + 107,56258 = 0,24514$ *a. u.* The first is more advantageous than the second one on $0,0622$ *a. u. (151 kJ)*.

Table 21. Some characteristics of atoms of the reactive center of *$(CH_3)_2NCH_2OOC(CH_3)_3$* **homolysis on the reaction way. Decomposition upon** *–O–O–* **bond** *($\ell_{C-H} = 1,095$)*

№	$\ell_{O-O}, \overset{0}{A}$	$-q_1$	q_2	$-q_3$	$-q_4$	$-q_5$	$-q_6$	q_7	$-E_{tot}$, a. e.	μ, D
1	1,49	0,2026	0,3549	0,0629	0,0625	0,1959	0,1800	0,2291	1087,56258	1,44
2	1,79	0,1928	0,3486	0,0575	0,0575	0,2017	0,2078	0,2278	107,41493	1,56
3	1,90	0,1901	0,3468	0,0558	0,0556	0,2025	0,2159	0,2262	107,3797	1,65
Synchronous decomposition upon O–O and C–H bonds										
2	1,79 $\ell_{C_2-H_3} = 1,295$	0,1900	0,3492	0,0510	0,0812	0,1182	0,2064	0,2270	107,38760	1,54
3	1,90 $\ell_{C_2-H_3} = 1,395$	0,1855	0,3465	0,0454	0,0918	0,1996	0,2147	0,2262	107,31744	1,65

3.6. PECULIARITIES OF ELECTRON STRUCTURE OF PEROXIDE ETHERS

This *Chapter* is distinguished explicitly, since, from our point of view, it is enough important at the following interpretation of the peresters reactivity in different reactions of their chemical transformation. It is necessary to note, that the set forth below investigations have not the partial character, concerning to only the peresters, but are also by general regularities for a whole series of the organic peroxides *(see* Chapter 3.11 *«About classification of organic peroxides»).*

It was informed earlier, that the *cis*–conformation (position of the carbonyl group with respect to the peroxy one) is typical for the peresters. Let us consider this question more in detail from the positions of the electron structure of peroxides.

On Figure 35 it is presented the distribution of charges in *cis*– and *trans*–conformers of *tert.*–butylperacetate and in *cis*–conformer of *tert.*–butylperoxytrime-thylacetate. The charge on β–oxygen is considerably less in *cis*–conformer. The change of the methyl radical on the *tert.*–butyl one manifests itself mainly on the charge of the carbonyl atom of the oxygen. The charges of oxygen atoms of peroxide bridge in these peresters are practically the same. From this it follows logically to suppose that the distribution of charges on the reactive center into peresters is caused both by the inductive influence of the substitutes and by the position of carbonyl group *(cis–, trans–)* with respect to the peroxy one, *i. e.* here the intramolecular interaction between O_β and carbonyl oxygen should be take place.

The analysis of *MO* showed, that for the peresters there is a series of *MO*, in which the compliance of the *AO* blades of carbonyl atom of oxygen and of the β–peroxy one is observed. The most clear such picture is traced for the stereo–determining *MO*, which has the maximal endowment into the interaction of O_β–oxygen of carbonyl [73].

An analysis of the own vectors of presented *MO* permits to suppose, that it corresponds to the *p*–unshared pair of electrons of the carbonyl atom of oxygen. Unshared pair of electrons of the carbonyl atom is placed on the *M* level of *AO*, *M* of *AO* takes part into *O–O* bond formation, *AO* is occupied by the electrons of the unshared pair O_β, and *N* of *AO* takes part into formation of

$\diagdown \atop \diagup$ $C{=}O$ bonds. Atoms O_γ and O_β are located in the same flatness and *AO M, M',* *N, N'* are overlapping between themselves in accordance with the σ–type.

3.7. AN INFLUENCE OF THE SUBSTITUENTS NATURE ON HOMOLYSIS OF PEROXY COMPOUNDS

A reacting ability of the molecules and others particles in reactions is often characterized by different tunes calculated in accordance with the *MO* (molecular orbitals) method *(namely,* by energy of the localization, polarizability of bonds, an order of bonds, free valency and *ect.*). Sometimes for the reaction of polar molecules it is enough restricted oneself to the component $q_r \Delta \alpha_r$ from a general equation of the perturbation theory (where q_r is a charge on the atom, and $\Delta \alpha_r$ is a change of the *Coulomb's* integral; $\Delta \alpha_r =$ *const* for the presented reacting series). This index (a charge) is often useful into reactions of the nucleophylic substitution.

From the other point of view, an organic chemistry is characterized by a wide application of different correlation equations by *Tafft, Hammet* and *ect.* type for obtaining the quantitative information about investigated processes. Classical *Hammet's* equation is as follow: $lgk/k_0 = \rho\sigma$, where k_0 is a rate of the standard compound, k is a rate of the sought one. The value σ was interpreted by *Hammet* as a measure function of the charge on reacting center transmitted by demonstration of the inductive effect, and ρ was interpreted as the sensitivity of reaction to the change of a charge on reacting center.

Such interpretation of σ value predetermined the following substantiation of the correlation ratios as a searching of the dependencies between σ–constants and calculated values, at first, charges and, afterwards, other indexes of the reacting ability.

Hence, an analysis of the charges on the atoms gives some qualitative conception about reacting ability of the molecules.

In our particular case the analysis of the charges distribution on the atoms of the reacting center of peroxides *(ROOR)* for *cis–* and *trans–*conformation means, that the redistribution of charges proceeds at the expense of the interaction of valence uncoupled atoms; in other words, this supposes the dependence of the peroxides homolysis activation energy also on the value of energy of such interaction, since as a criterion of such interaction can be accepted the total orbital energy of pair–wise interaction of valence uncoupled atoms of reacting center.

Let us consider from this point of view more in detail the influence of the substituents nature on electron structure of peroxides and on the process of their homolytic decomposition upon $-O-O-$ bond.

Under calculations of the investigated perethers both into equilibrium and transition states the geometry was assumed as same as in a case of methylperacetate. At this, it was taken into account the geometric parameters of the substituents.

A distribution of charges on the atoms of the molecules is represented in Table 22.

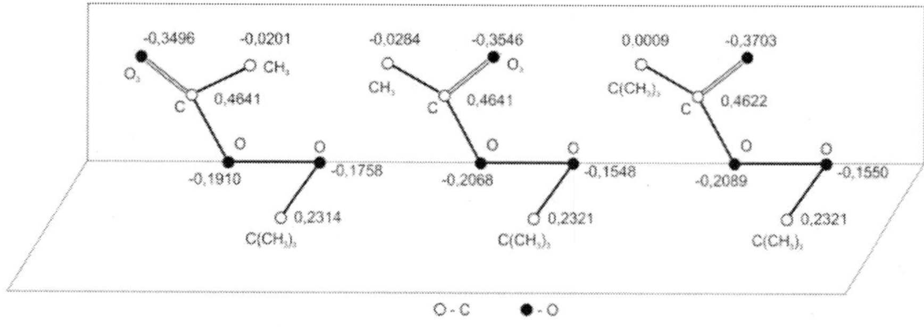

Figure 35. The distribution of charges in *cis–* and *trans–*conformers of *tert.–*butylperacetate and *cis–tert.–*butylperoxytrimethylacetate.

Table 22. A distribution of charges upon atoms for reactive center of perethers

№	$R-C^{O^2}\!\!-O\,O\,C(CH_3)_3$	$\ell_{O-O}=1{,}49\,\overset{0}{A}$				$\ell_{O...O}=1{,}85\,\overset{0}{A}$			
		q_{C1}	$-q_{O2}$	$-q_{O3}$	$-q_{O4}$	q_{C1}	$-q_{O2}$	$-q_{O3}$	$-q_{O4}$
1	CH_3	0,4641	0,3546	0,2068	0,1548	0,4698	0,3497	0,1765	0,2066
2	C_2H_5	0,4599	0,3555	0,2058	0,1556	0,4637	0,3517	0,1792	0,2066
3	$n{-}C_3H_7$	0,4549	0,3579	0,2665	0,1568	0,4577	0,3557	0,1833	0,2062
4	$n{-}C_4H_9$	0,4560	0,3565	0,2061	0,1560	0,4592	0,3541	0,1818	0,2064
5	$iso{-}C_3H_7$	0,4549	0,3579	0,2065	0,1568	0,4577	0,3559	0,1833	0,2062
6	$tert{-}C_4H_9$	0,4622	0,3703	0,2089	0,1550	0,4716	0,3743	0,1850	0,2062
7	CH_2Cl	0,4765	0,3476	0,1971	0,1569	0,4820	0,3530	0,1710	0,2087
8	CCl_3	0,4665	0,3095	0,1886	−0,1422	0,4720	0,3145	0,1585	0,2040
9	Cl	0,5053	0,2754	0,1766	0,1422	0,5103	0,2721	0,1425	0,2089
10	C_6H_5	0,4561	0,3613	0,2086	0,1592	0,4631	0,3630	0,1896	0,2061
11	$CH_2{=}CH$	0,4509	0,3523	0,2125	0,1547	0,4659	0,3512	0,1846	0,2062
12	$CH_2{=}C(CH_3)$	0,4431	0,3405	0,2031	0,1548	0,4579	0,3392	0,1772	0,2062

It can be seen from the data of Table 22 that a series of substitutes which are differed upon their donor–acceptor properties have a different influence on the distribution of charges upon atoms of reactive center of peresters both into equilibrium and transition states.

For a series of normal alkyl substitutes the positive charge on carbonyl carbon C_1 is expected decreased with their electron–donor properties increasing whereas the negative charge is increased on the oxygen of carbonyl.

Let us suppose that in presented case there are observed the classical conceptions about donor–acceptor influence of substitutes. However, an increasing of the electron–donor properties into normal range of substitutes should lead to the increase of the negative charge also on oxygens of the peroxide bridge $(O_\alpha{-}O_\beta)$ as a result of the sufficiently great electronegative of its atoms. As it can be seen from the Table 22 the charge on O_α is not only increased but also on the contrary is decreased. It is follow from this that the general conceptions about electronic properties of the substitutes (inductive, mesomeric and *ect.*) can not be the starting point for the explanation of the perethers structure and their reactive ability. It can be observed some correlation between any characteristics of substitutes (σ–constants, q–charges of carbonyl carbon atom) and their homolysis energy activation. However, such correlations are extremely limitary since they most likely reflect the definite type of the interaction of substitute with the reactive center.

If to issue from the conceptions about the structure of peroxy ethers then their reactive ability into homolysis reactions can be easy interpreted from the point of view of substitute nature influence.

Firstly, it should be supposed that the all perethers into equilibrium state have the *cis*–structure (a location of the carbonyl group respectively to the peroxy one). Secondly, it is assumed that the carbonyl oxygen (O_γ) interacts with the α– and β–oxygen of the peroxy bridge and C_1 interacts with O_β. Thirdly, it is suspected that the activation energy of the homolysis reaction is also caused by these interactions.

Starting from these points of view it can be easy explained the influence of the substitutes' nature on the reactive ability of the perethers into homolysis reactions.

Into activated complex the positive charge on C_1 and all oxygen atoms of peroxide system is increased in a case of the normal alkyl substitutes at the internuclear distance $-O-O-$ increasing *(see* Table 22*)*. Increase of the charge on C_1 can be explained at the expense of weakening the non–valency bond $C_1...O_\beta$.

Interactions of $O_\gamma...O_\beta$ and $C_1...O_\beta$ into activated complex is weaker. Weakening of the $O_\alpha...O_\beta$ bond leads to some delocalization of charge upon the all reactive center. The charges upon O_γ and O_α are decreased and upon O_β are increased. The last is explained by strong weakening of the interactions $O_\gamma...O_\beta$ and $C_1...O_\beta$ in consequence of the charge localization on the fragment $...OC(CH_3)_3$. The charges reducing upon O_α and O_γ would be explained at the expense of unpaired electron delocalization upon the fragment of complex

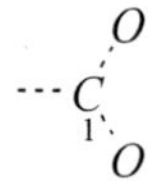

$$...\overset{\displaystyle O}{\underset{\displaystyle O}{C_1}}$$

in connection with the tendency towards the alignment of C_1-O_γ and C_1-O_α bonds. However, the charge increase on a C_1 into activated complex contradicts to this assumption. If the alignment of the electron density on the oxygen atoms of O_γ and O_α took place only upon the chemical bonds then in any case the positive charge on a C_1 wasn't decreased. Probably, in order to explain this fact it should be assumed that the alignment of the charges on a O_α and O_γ proceeds via their triaxial (space) interaction, that is via the formation of some bond $O_\gamma...O_\alpha$.

Table 23. Full electron energies for the peresters into initial and transition states

№	$R-\overset{O^2}{\underset{1\ \ 3\ \ 4}{C}}-O\,O\,C(CH_3)_3$	$\ell_{O-O} = 1{,}49\,\overset{0}{A}$, $-E_{1tot}$, a. u.	$\ell_{O...O} = 1{,}85\,\overset{0}{A}$, $-E_{2tot}$, a. u.	$-E_{2tot}-E_{1tot}$, a. u.	$E_{act.}$, kJ/mole
1	CH_3	103,0953	102,9033	0,1920	159,0
2	C_6H_5	138,8176	138,6284	0,1892	150,6
3	C_2H_5	111,5322	111,3434	0,1888	146,0
4	CH_2Cl	118,5089	118,3200	0,1889	149,4
5	$CH_2=CH$	109,7456	109,5570	0,1886	—
6	$CH_2=CH(CH)_3$	118,1863	117,9983	0,1880	136,8
7	iso–C_3H_7	119,9691	119,7813	0,1878	133,1
8	tert–C_4H_9	128,3937	128,2080	0,1857	128,0
9	CCl_3	149,3622	149,1769	0,1853	126,8

Isoalkyl substitutes *(see* Table 22*)* although are more stronger donors in comparison with their normal analogues, however the positive charge on a C_1 is even some higher than in a case of the normal substitutes. First of all, this is caused by sufficiently strong visualization of non–valency interaction effect between the substitute R and O_γ–oxygen.

In a case of the chloride–substituted peresters the charge on a C_1 is increased; respectively, the molecules on the atoms of oxygen are decreased. It is clear, that such distribution of charges on the atoms is caused by sufficiently great electronegativeness of the chlorine atoms. The strength of a $-O-O-$ bond is decreased with the number of chlorine atoms increasing.

Unsaturated substitutes C_6H_5, $CH_2=CH-$, $CH_2=C(CH_3)$ decrease the charge on a C_1 till the value even less than the alkyl substitutes. Most likely this is explained by increasing the interaction between the substitute and carbonyl atom of oxygen via space.

In Table 23 there are data concerning to the full electron energies for the peresters into initial and transition states.

As we can see from the above presented Table 23, the calculated "activation energies" $(E_1 - E_2)$ are very far from their experimental values. However, it is necessary to notify, that the difference of diapason for calculated "activation energies" is near to the experimental difference – 33 *kJ/mole.*

Table 24. Distribution of the charges upon atoms of the reactive centre for diakyl peroxides into equilibrium and activated states

№	Peroxide	$\ell_{O_2-O_3}=1{,}49\overset{0}{A}$				$\ell_{O_2...O_3}=1{,}85\overset{0}{A}$			
		q_{C1}	$-q_{O2}$	$-q_{O3}$	q_{C4}	q_{C1}	$-q_{O2}$	$-q_{O3}$	q_{C4}
1	$H_3\underset{1}{C}-\underset{2}{O}\underset{3}{O}-\underset{4}{C}H_3$	0,2600	0,1537	0,1537	0,2600	0,2513	0,1714	0,1714	0,2513
2	$C_2H_5OOC_2H_5$	0,2524	0,1674	0,1674	0,2524	0,2425	0,1870	0,1870	0,2404
3	$n-C_3H_7OOC_3H_7$	0,2484	0,1723	0,1723	0,2484	0,2404	0,1922	0,1922	0,2404
4	$n-C_4H_9OOC_4H_9$	0,2452	0,1732	0,1732	0,2452	0,2372	0,1930	0,1930	0,2372

From this it follows that the experimental activation energies for presented series of investigated peresters are caused by their electronic structure and the contribution of space component of energy and also the influence of the solvent nature into a process are insignificant.

Distribution of the charges on a fragment $C_1-O_2-O_3-C_4$ for dialkyl peroxides into initial and transition states are represented in Table 24.

From the data presented in Table 24 it can be seen that the increasing of donor properties of substitutes in a range of the peroxides 1–4 increases the charge on the oxygens of bridge. Into peroxide state the charges on the oxygens are some increased as a result of steaming the electrons forming the – O–O– bond. Delocalization of electrons facilitates to lowering into activated state even charges on the carbon atoms of substitutes *(see* Table 24*)*.

Summarizing the above–said it can be done the following generalizations: firstly, the activated complex forming via the peresters homolysis process by its electron structure is on the whole different from any starting reagent and final products. Secondly, this complex can be represented into the first assumption by two autonomous fragments (*I*) and (*II*):

$$R\ldots\overset{\ldots\overset{O}{}}{C}\ldots O\cdots O\ldots C\overset{/}{\underset{\backslash}{}}-$$

$$\mathbf{I} \qquad\qquad \mathbf{II}$$

The fragment (*II*) is near to free radical $\overset{\bullet}{O}C\overset{/}{\underset{\backslash}{}}-$ upon its electron characteristics. Thirdly, an influence of the substitutes on reactive ability of the peresters into reactions of their homolysis is displayed accordingly to the complicated mechanism leading to the change of the electron density upon the all reactive center. Mechanisms of the substitutes influence on the electron structure of peresters into equilibrium and activated states are like and are realized multicenterly both upon valency bonds and non–valency via the space. The difference is only fact that the less weaker non–valency and valency interactions of atoms into equilibrium state are intensified into the activated complex and the strong non–valency and valency interactions of atoms into equilibrium state on the contrary are relaxed.

3.8. A ROLE OF THE UNSHARED PAIRS INTO PEROXIDES HOMOLYSIS REACTIONS

In *References* there are forcible arguments of fact that the unshared electron pairs of the neighboring atoms can interact in such a way, that physical properties of the considered molecule are changed; *for example*, the ionization potential is lowered, the *UV*–spectrum is changed and *ect*. The lapping of the unshared electron pairs orbitals into molecule leads to the light adsorption in more long–wave field.

The influence of unshared electron pairs is exhibited also on chemical properties. Undoubtedly, the stability of these or others organic peroxides in a great degree should be depend on the interaction of unshared electron pairs of the peroxide bridge with the fragments of peroxides. Therefore, this interaction in some manner should be reflected on the reactive ability of peroxy compounds in different reactions, and first of all, into reactions of homolytical decomposition.

The analysis of proper vectors and proper values of *MO* for peroxides and oxy–radicals showed that the electrons of n–pair of oxygen atoms occupy the boundary and next below occupied *MO* (it is observed the maximal probability C^2). This is enough clear observed into oxy–radicals. Thus, into *OH* radical $C^2 = 1$ on the boundary occupied *MO* both upon α– and β–spin. Into CH_3O $C^2 = 0,6 - 0,7$ on corresponding *MO*.

If the *AO* of n–pairs of oxygen atoms do not overlapped with others *AO*, then its occupation is equal to 2,0 as same as into a case of the hydroxyl radical. But, as it follows from the Table 25, *AO* orbital of n–pair of spatially branched radicals and molecules of peroxides are overlapped with a range of *AO* atoms of system. And, hence, the electrons of n–pair are delocalized into the space and contribute the definite share into the bond strength either valency or valency unbound atoms. The degree of electrons delocalization of n–pair into space is the occupation of *AO* of n–pair (see Table 26).

Table 25. Integrals of *AO* overlapping for *n*–pair with *AO* of other atoms into peroxides and radicals

			H_1	O_2				O_3				H_4
			S	S	P_x	P_y	P_z	S	P_x	P_y	P_z	S
O	P_x	$\dot{O}H$	0	0	1	0	0	—	—	—	—	—
		H_2O_2	0	0	1	0	0	0	0,1408	0	0	0,0610

			H_1	H_2	H_3	C_4			
			S	S	S	S	P_x	P_y	P_z
O	P_x	$CH_3\dot{O}$	−0,0359	0,0359	0	0	0,1613	0	0
		CH_3OOCH_3	−0,0327	0,0327	0	0	0,1568	0	0

			O_5				O_6			
			S	P_x	P_y	P_z	S	P_x	P_y	P_z
O	P_x	$CH_3\dot{O}$	—	—	—	—	—	—	—	—
		CH_3OOCH_3	0	1	0	0	0	0,1362	0	0

			C_7				H_8	H_9	H_{10}
			S	P_x	P_y	P_z	S	S	S
O	P_x	$CH_3\dot{O}$	—	—	—	—	—	—	—
		CH_3OOCH_3	0,0672	−0,0434	−0,0887	0	0,0058	0,0380	0,0380

Table 26. Occupations and proper vectors of radicals and molecules

№		P–AO	Spin	1	2	3	4	5
1	$\dot{O}H$		α	0	0	0	—	—
			β	0	0	1	—	—
2	H_2O_2	1,9736	—	0,0164	−0,0693	0,3424	0,4447	−0,3181
3	$CH_3\dot{O}$	1,9581	α	0	0	0	0,5878	0
			β	0	0	−0,5501	0	0
4	CH_3OOCH_3	1,9497	—	0,0429	−0,1004	0,1276	−0,1453	0,3176

№		P–AO	Spin	6	7	8	9	10
1	$\overset{\bullet}{O}H$		α	—	—	—	—	—
			β	—	—	—	—	—
2	H_2O_2	1,9736	—	0,5386	0,5248	—	—	—
3	$CH_3\overset{\bullet}{O}$	1,9581	α	0	0,7963	—	—	—
			β	–8221	—	—	—	—
4	CH_3OOCH_3	1,9497	—	0,3788	–0,1965	–0,1238	0,0952	–0,0984

№		P–AO	Spin	11	12	13
1	$\overset{\bullet}{O}H$		α	—	—	—
			β	—	—	—
2	H_2O_2	1,9736	—	—	—	—
3	$CH_3\overset{\bullet}{O}$	1,9581	α	—	—	—
			β	—	—	—
4	CH_3OOCH_3	1,9497	—	0,0925	0,5531	0,5422

If into OH radical the AO occupation of n–pair is equal to 2, then into CH_3O is equal to 1,9581, where AO–pairs are overlapped with $S–AO$ of two protons symmetrically located relatively to $C–O$ bond.

In molecules of peroxides the AO of n–pairs of one oxygen are overlapped with the AO of other oxygen taking part into formation of $O–X$ bond ($X = H$, group of atoms) accordingly to σ type and also with a series of corresponding to AO groups X (see Table 26).

Hence, it becomes clear the role of the unshared pairs contribution into the strength of the peroxide bond. Naturally, that the role of the n–pairs via the process proceeding is changed.

In order to estimate the contribution of n–pairs into the strength of the peroxide bond it is necessary to determine their role both into the starting molecules and into the corresponding radicals. This difference can characterize their contribution into the strength of the $–O–O–$ bond or the activation energy.

Since in the peroxides the unshared n–pairs of oxygen of peroxide bridge are some delocalized upon the system, then the participation degree of AO of n–pair in others MO is determined by the square of coefficients at AO of the unshared pair. Effective energy of the molecular «localized» orbital of the unshared pair can be represented by the following expression:

$$E_{eff.} = c_{i1}^2 E_1 + c_{i2}^2 E_2 + ... + c_{ig}^2 E_g \qquad (6)$$

where c_{ig} is the proper vectors of i–AO into g–MO, E_g is the energy of g–MO.

We suppose that the difference $E_{eff.}$ for the unshared pairs of oxygen bridge of peroxides and $E_{eff.}$ for the unshared pairs of radicals (products of the peroxides decomposition) can serve by some characteristic of their influence on stability of the peroxide bond and activation energy of the homolysis process.

Really, as we can see from the data of Table 27, it is observed some tendency to the cymbate change of this difference for effective energies of MO of unshared pairs for dialkyl peroxides with their activation energy of homolytical decomposition.

Table 27. The values of effective energies of delocalized unshared pairs for oxygen atoms of peroxide bridge and for oxygen atom of alkoxyl radicals

№	Peroxide	$E_{1eff.}$, a. u.	$E_{2eff.}$, a. u.	$E_{2eff.} - E_{1eff.}$, a. u.	$E_{act.}$, kJ/mole
1	H_2O_2	–2,4600	–2,5654	0,1054	201
2	$(CH_3)_3COOC(CH_3)_3$	–2,3224	–2,4104	0,0880	159
3	CH_3OOCH_3	–2,4116	–2,4988	0,0872	155
4	$C_2H_5OOC_2H_5$	–2,3736	–2,4552	0,0816	146
5	$(CH_3)_2NCH_2OOC(CH_3)_3$	–2,4459	–2,4473	0,0024	134
6	$CH_3\overset{\displaystyle O}{\overset{\|}{C}}$ - $OO\ C(CH_3)_3$	–2,5080	–2,3508	0,1572	159
7	$C_6H_5\overset{\displaystyle O}{\overset{\|}{C}}$ - $OO\ C(CH_3)_3$	–2,4710	–2,2250	0,2460	150
8	$CH_3\overset{\displaystyle O}{\overset{\|}{C}}$ - $OO\ CH_3$	–2,6366	–2,3846	0,2520	134
9	$(CH_3)_3C\overset{\displaystyle O}{\overset{\|}{C}}$ - $OO\ C(CH_3)_3$	–2,5078	–2,2317	0,2761	128

Effective energy $(E_{2eff.})$ of localized *MO* for unshared *n*–pairs of oxygen atoms for dialkyl peroxides is some higher then the total characteristic $E_{1eff.}$ for the corresponding radicals. It follows from this that the delocalization of pairs in the presented peroxides is higher than into the corresponding radicals; this means that with the degree delocalization increasing the probability of the –O–O– bond loosening is lowered. For the alkoxyradicals the delocalization degree of *n*–pairs is lowered in a range of their activity decreasing: *OH* > *CH₃O* > *C₂H₅O* > *(CH₃)₃CO*.

Really, the more delocalization degree of unshared pairs for oxygen of dialkyl peroxides and corresponding radicals forming under its decomposition the more should be the strength of the broken –O–O– bond, and, respectively also the energy of the process activation.

Some other picture is observed in this aspect for peresters. As we can see from the data of Table 27, the effective energy of the unshared pairs $E_{1eff.}$ for the forming radicals is some higher than for corresponding molecules; that is why the negative values of $E_{2eff.} - E_{1eff.}$ are obtained.

3.9. ABOUT CORRELATION BETWEEN THE ENTHALPIES OF PEROXY COMPOUNDS FORMATION INTO EQUILIBRIUM AND TRANSITION STATES INTO REACTIONS OF THEIR HOMOLYTICAL DECOMPOSITION

The thermal effect *(Q)* of the peroxy compounds decomposition reaction can be represented by the following expression:

$$Q = E_1 - E_2 \tag{7}$$

where E_1 and E_2 are the activation energies of direct and reverse reactions. If to consider, that $E_2 \approx 0$, then $Q \approx E_1$; that is the heat of the dissociation is equal to the activation energy of direct reaction. The value 0 can be calculated based on the thermo–chemical data (heat of the reagents and the reaction's products formation).

If for the reagents (peroxides) ΔH can be measured and calculated, then for the products of reaction this value (alkoxyl radicals) is determined in a very complicated manner.

However, $E_2 \neq 0$, and for the reaction of radicals recombination consists of 21 *kJ/mole* [74]. Then the activation energy can not be estimated in accordance with the above–said arguments.

In *Ref.* [75] it was shown by us, that between free energy of activated bond into transition state *(ΔF_{freeK})* and free energy of the broken bond into the starting reagent *(ΔF_{freeBr})* the linear dependence exists for the organic substances homolysis monomolecular reactions:

$$\Delta F_{freeK} = \alpha \Delta F_{freeBr} + B \tag{8}$$

where $\alpha \approx 2$, $B \approx 7$.

This equation permits to estimate the free activation energy of monomolecular reaction decomposition for a series of organic substances upon bonds $C–C$, $C–H$, $C–X$ (X are heteroatoms or heterogroups).

To check of this correlation for peroxy compounds is impossible, since to estimate the free energy of $–O–O–$ bond in peroxides is complicate; at the same time, to apply the additive schemes for the thermodynamic characteristics calculations is practically not suitable.

Taking into account the fact, that the internuclear distance of $–O–O–$ bond for the peroxides into activated complex is elongated on a small–scale magnitude, it can be expected of some geometric likeness between this complex and equilibrium state. And, if that's the case, logically to suppose that also some properties of the activated complex are intermediate between the properties of reagents and the products despite the fact, that the free energy of the activated complex has the extreme value.

For example, it can be supposed for peroxy compounds that between the enthalpies of the activated complex formation (the sum of the starting reagent enthalpy and activation enthalpy) and the enthalpies of the starting reagents into reactions of the monomolecular decomposition the some definite dependence should be existed. Such dependence is represented on Figure 36.

However, it is necessary to take into account, that on the *Figure* 36 a not great quantity of points is represented, although the diapason of heat activation change is sufficiently wide; that is why, this dependence is most likely qualitative. The last first of all is caused by the absence in *Ref.* the heat formation of peroxy compounds. Very likely, that the correlations by similar type will be useful in practice.

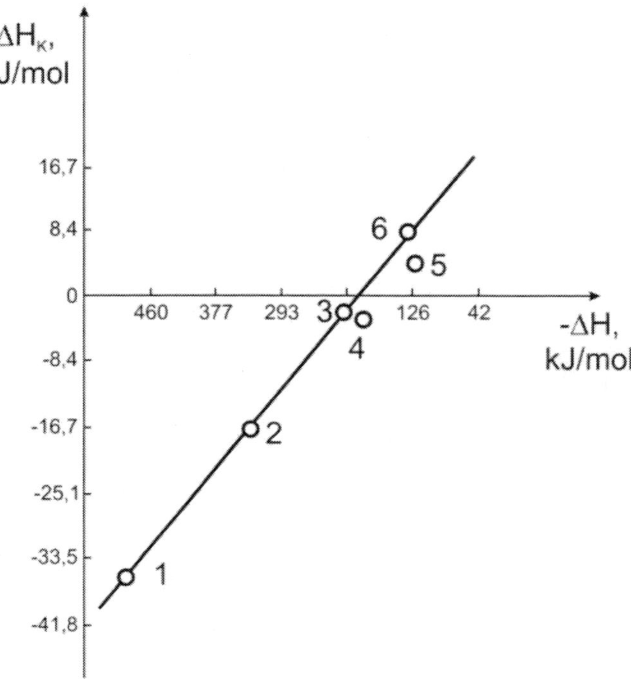

Figure 36. Dependence of the enthalpy of activated complex formation *(ΔH_K)* on the enthalpy of starting reagent formation *(ΔH)* at [*76*] for homolysis reaction of

peroxides: *1. $CH_3 \overset{O}{\overset{\|}{C}} - OO\ CH_3$; 2. $(CH_3)_3COOC(CH_3)_3$; 3. CH_3COOH; 4. $C_2H_5OOC_2H_5$; 5. CH_3OOCH_3; 6. H_2O_2.*

3.10. ABOUT PREEXPONENTIAL FACTORS OF THE HOMOLYSIS REACTIONS OF PEROXY COMPOUNDS

In most cases a decomposition of the molecules into two radicals proceeds via semi–rigid activated complex [77]. Under formation of activated complex thee are possible the following changes of the internal freedom degrees:

i)   the decrease of the deformative oscillation frequency (in 2 times);
ii)   the disengagement of the internal rotation relatively the broken bond, if this rotation was delayed into equilibrium state;

iii) the change of rotatory oscillation frequencies relatively the bonds that is the change of rotatory moving of R groups neighbouring to the reactive center [77].

In a case of peroxy compounds under the broken of $-O-O-$ bond the barrier of the internal rotatory of R groups should be decreased, that is here obviously in some manner the effect of the neighbouring groups should be showed. The attempts to explain the high preexponentials under peroxides decomposition without taking into account of the R rotatory were unsuccessful [77].

It is generally known, that if the activated complex has a great size or the «incoherent» structure, then the activation entropy is more zero, and hence, it can be expected the anomaly high value of the preexponential. If the activated complex is more ordered (stringent) than the equilibrium state, then it can be expected the anomaly low value of the preexponential.

Accordingly to *Benson* [78] the decomposition of the great molecules should be accompanied by visible increasing the translational entropy, and consequently, for them it should be characteristic the higher values of the preexponential. This can be explained by the bond extension resulting to the decrease of the power constants characterizing the four pendular oscillations of groups relatively to the bond line.

In accordance with *Benson* [78] the deformative oscillations frequency into activated complex can be decreased in 5–6 times. Such decreasing per se corresponds to the free rotation of groups into activated complex.

Homolytical decomposition of the organic peroxides upon $-O-O-$ bonds depending on the structure is characterized by a wide diapason of the preexponential factors values $(10^{13} - 10^{18} \ s^{-1})$. We will attempt to explain the low and high values of the preexponentials for the monomolecular reaction of the peroxides decomposition from our point of view.

Let us estimate the activation entropy for the reaction of homolytical decomposition of peroxide by structure *(CH₃)₂NCH₂OOC(CH₃)₃*. For this compound the preexponential is equal to $10^{14,5} \ s^{-1}$; this corresponds to the activation entropy *(ΔS⁺)* at 393 K 56,3 *e. u.* The spin endowment into the reaction activation entropy is equal to zero. The endowment of the symmetry into $\Delta S^{\neq}$ for presented system is also equal to zero. The entropy of the internal rotation endowments into $\Delta S^{\neq}$ will not take into account also.

In Table 28 there are oscillation endowments into $\Delta S^{\neq}$ for the reaction:

$$(CH_3)_2NCH_2OOC(CH_3)_3 \rightarrow (CH_3)_2NCH_2O + OC(CH_3)_3 \qquad (9)$$

Table 28. The oscillation endowments into $\Delta S^{\neq}$ at 400 K for the reaction $(CH_3)_2NCH_2OOC(CH_3)_3 \rightarrow (CH_3)_2NCH_2O + OC(CH_3)_3$

№	Endowments	$\Delta S^{\neq}$, e. u.
1	Valency oscillation –O–O– (895 sm^{-1}) reaction coordinate	–1,68
2	Valency oscillations:	
	C–O (1076 → 764) sm^{-1}	0,42
	O–C (1200 → 710) sm^{-1}	2,1
	C–N (1055 → 675) sm^{-1}	1,29
	N–C (1035 → 675) sm^{-1}	1,29
	(C–H) (3100 → 2200) sm^{-1}	
	(C–C) (1000 → 675) sm^{-1}	1,29
3	Deformative oscillations:	
	2OCH (965 → 704) sm^{-1}	4,62
	2COO (252 → 150) sm^{-1}	5,46
	CO (630 → 400) sm^{-1}	4,62
	2CH (1370 → 980) sm^{-1}	1,68
	3OCC (400 → 280) sm^{-1}	5,46
	2CC (400 → 300) sm^{-1}	5,46
	3CCC (420 → 300) sm^{-1}	5,46
	Full change of the activation entropy	76,4

The frequencies of valency and deformative oscillations into equilibrium state were taken from the Table 3. Into activated complex the decreasing of the deformative frequencies were assumed in 2 times less and valency ones – in 1,5 times lower than into the initial state [78]. Oscillation entropies were estimated in accordance with the [78].

It can be seen from the Table 28 that the calculated $\Delta S^{\neq}$ (76,4 $e.\ u.$) is disagreed with the experimental value $\Delta S^{\neq}$ (56,3 $e.\ u.$). Taking into account even the fact, that the preexponential is experimentally determined with a sufficient great error anyway is complete to compare at least qualitatively these two magnitudes.

If to assume, that into activated complex the frequencies of the deformative oscillations are decreased in 3 or 5 times, then the divergence between the estimated value $\Delta S^{\neq}$ and experimental one will be much great and to explain so great difference is impossible.

Let us consider the above–mentioned reaction with taken into account the peculiarities of the structure of *(CH₃)₂NCH₂OOC(CH₃)₃* compound. If to take into account that the reactive center of this compound has the cyclic structure

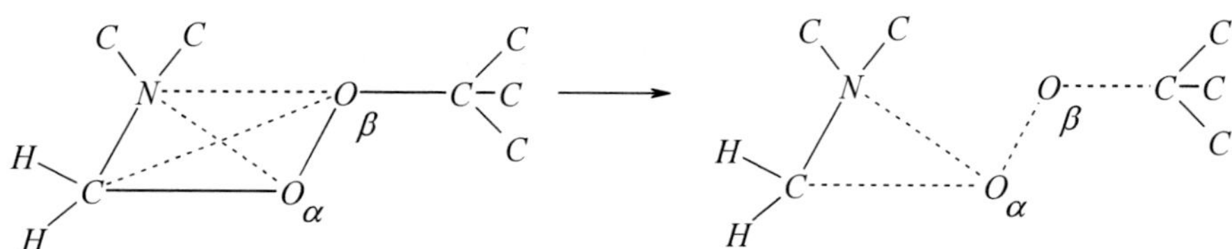

then it becomes clear some moments explaining the value of activation entropy of the decomposition reaction. Activated complex *(AC)* is stabilized mainly at the expense of the interaction $N...O_\alpha$. Others bonds $N...O_\beta$, $O_\alpha–O_\beta$ into *AC* are greatly weakened, that is why not the all deformative oscillations will gave the endowment into $\Delta S^{\neq}$ of the reaction. The deformative oscillations of *2COO*, *3OCC* and *3CCC* will gave the most essential endowments. Such oscillations as *2HCN*, *2HCO*, *2CNC* are less applied into activated complex as a result of fact that the nitrogen atom both in equilibrium and into *AC* state connected with the atoms of O_β and O_α respectively. Since the bonds *C–O* and *O–C* into *AC* are shortened and are neared to the bonds in the free radicals *(CH₃)₂NCH₂O•* and *•OC(CH₃)₃*, then it can be expected the decrease of the deformative oscillations *COO*, *OCC* and *CCC*. Obviously, it is necessary to take into account also the decrease of the deformative oscillation CNO_α into *AC*. Therefore, $\Delta S^{\neq}_{oscill.}$ is equal to ≈ 48,3 *e. u.* Taking into account that $\Delta S^{\neq}_{rot.}$

$$\approx R\ln\frac{r_{\neq}^2}{r^2} \approx 6 \; e.u., \; r = 3,82 \; \overset{0}{A}; \; r_{\neq} = 4,02 \; \overset{0}{A},$$ we will obtain $\Delta S^{\neq} \approx 54,3$ *e. u.*, that

is satisfactory agreed with the experimental value 56,6 *e. u.*

Let us consider other example of the entropy estimation for the reaction of *tert.–*butylperbenzoate thermal decomposition.

Under calculation of $\Delta S^{\neq}$ it was taken into account only the freedom degrees which do the main endow into the activation entropy (Table *s* 1, 2). Oscillation endows of the deformative oscillations of *HCC* and valency *C–H* for methyl groups were also neglected, since they are not changed via the reaction process and don't of any endow into $\Delta S^{\neq}$. Also it were not taken into account endows of the valency *C–H* and deformative *HCC* for the benzene ring (*see* Table 29).

Table 29. Oscillation endows into activation entropy of the reaction

$$C_6H_5\overset{O}{C} - OO\,C(CH_3)_3 \longrightarrow C_6H_5\overset{O}{C} - O + O\,C(CH_3)_3 \text{ at } 400\ K$$

	Endowments	$\Delta S^{\neq}$, a. u.
Valency oscillations (sm^{-1})		
1	$O–O$ (933) $\rightarrow$ coordinate of reaction	−1,68
2	$_8C–C$ (1434 $\rightarrow$ 990)	0,84
3	$_6C–O$ (1245 $\rightarrow$ 830)	1,64
4	$_7C{=}O$ (1755 $\rightarrow$ 1400)	0,42
5	$_4O–C$ (1178 $\rightarrow$ 780)	1,64
6	$_1C–C$ (1186 $\rightarrow$ 780)	1,64
7	$_3C–C$ (1315 $\rightarrow$ 873)	1,29
8	$_2C–C$ (750 $\rightarrow$ 500)	2,52
9	$_9C–C$ (1280 $\rightarrow$ 853)	1,64
10	$_{10}C–C$ (1280 $\rightarrow$ 853)	1,64
11	$_{11}C–C$ (1280 $\rightarrow$ 853)	1,64
12	$_{12}C–C$ (1340 $\rightarrow$ 900)	1,29
13	$_{13}C–C$ (1315 $\rightarrow$ 876)	1,29
14	$_{14}C–C$ (1645 $\rightarrow$ 1100)	0,84
Deformative oscillations (sm^{-1})		
1	$_{7\,8}CCO$ (578 $\rightarrow$ 413)	4,62
2	$_{9\,6}OCC$ (578 $\rightarrow$ 413)	4,62
3	$_{6\,7}OCO$ (801 $\rightarrow$ 572)	3,78
4	$_{5\,6}OOC$ (219 $\rightarrow$ 156)	5,46
5	$_{4\,5}COO$ (171 $\rightarrow$ 122)	4, 20
6	$_{2\,4}CCO$ (475 $\rightarrow$ 340)	4,62
7	$_{1\,4}CCO$ (415 $\rightarrow$ 296)	5,46
8	$_{3\,4}CCO$ (351 $\rightarrow$ 250)	5,46
9	$_{8\,9}CCC$ (114 $\rightarrow$ 81)	3,78
10	$_{8\,14}CCC$ (293 $\rightarrow$ 209)	5,88
11	$_{9\,10}CCC$ (1078 $\rightarrow$ 770)	3,36
12	$_{10\,11}CCC$ (513 $\rightarrow$ 366)	5,04
13	$_{11\,12}CCC$ (699 $\rightarrow$ 470)	3,78
14	$_{12\,13}CCC$ (669 $\rightarrow$ 470)	3,78
15	$_{13\,14}CCC$ (578 $\rightarrow$ 413)	4,62
Full change of the activation entropy 85,7		

With taken into account the above–said the activation entropy for *tert.–butylperbenzoate* decomposition is equal to $\approx$ 85,7 *e. u.*, whereas the experimental value of $\Delta S^{\neq}$ for this reaction is equal to 67,2 *e. u.* at 400 K under condition that lgA = 16. Comparison of the calculated and experimental magnitudes shows, that they are greatly differed. Via homolysis reaction the non–valency interactions of the *tert.–butylperbenzoate* molecules $O_\gamma...O_\beta$ and $C...O_\beta$ are weakened into activated complex and $O_\gamma...O_\alpha$ become some stronger:

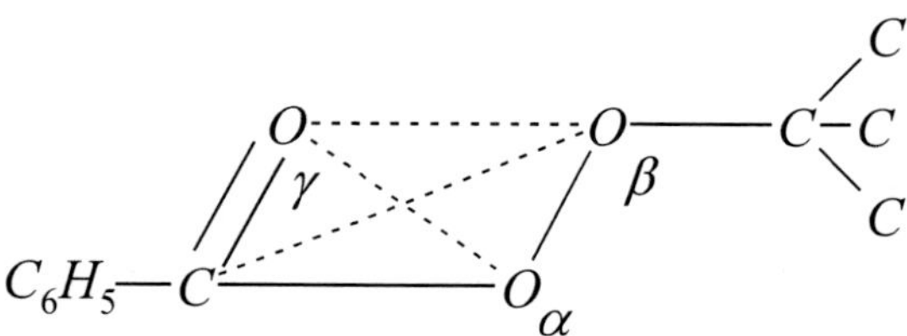

Consequently, the atom O_γ is sufficiently strongly bonded and deformative oscillation $O_\gamma CO_\alpha$ doesn't endow into $\Delta S^{\neq}$. Probably, that the deformative oscillations of the CCC benzene ring also are not changed into AC. So, with taken into account of the all above–said $\Delta S^{\neq}$ = 62,2 *e. u.* Adding to this value 6 *e. u.*, which take into account the rotation, we will obtain $\Delta S^{\neq}_{react.}$ = 68,2 *e. u.* that is in good agreement with the $\Delta S^{\neq}$ = 62,2 *e. u.*

Although the above–presented facts are qualitative, however they permit to do the following conclusion: it can be expected the activation entropy change for peroxy compounds via homolysis reaction depending on the O–O, $X...O_\alpha$, $C...O_\beta$ and $X...O_\beta$ bonds strength. The stronger these bonds, the rigid is equilibrium state, the less is its entropy, and this means that at more friability of the activated state the activation entropy will be cymbate to their strength. This conclusion is confirmed by the data presented in Table 30.

In *Chapter* 3.11 will be shown, that the activation energy of the peroxy compounds homolytic reaction is cymbate to the stabilization energy upon bonds O_α–O_β, $X...O_\alpha$, $X...O_\beta$ and $C...O_\beta$.

$$E_{act.} = \alpha E_{stab.} + B \tag{11}$$

The data of Table 30 pointed on *Figure* 37 into coordinates on $E_{stab.}$ is described by the dependence *(see Figure 37).*

Table 30. Dependence of the kinetic parameters of peroxides thermal decomposition on energy stabilization value

№	Reagent	lgA	$E_{act.}$, kJ/mole	$-E_{stab.}$, a. u.
1	$H_2\overset{\displaystyle\diagup OCH_3}{C}\text{ - }OO\,C(CH_3)_3$	18,00	175,7	1,0518
2	$CH_3\overset{\displaystyle\diagup O}{C}\text{ - }OO\,C(CH_3)_3$	17,80	150,0	1,0056
3	$C_6H_5\overset{\displaystyle\diagup O}{C}\text{ - }OO\,C(CH_3)_3$	16,00	150,6	0,9648
4	$CH_2Cl\overset{\displaystyle\diagup O}{C}\text{ - }OO\,C(CH_3)_3$	15,88	148,5	0,9604
5	$C_2H_5\overset{\displaystyle\diagup O}{C}\text{ - }OO\,C(CH_3)_3$	15,68	146,0	0,9554
6	$(CH_3)_2NCH_2OOC(CH_3)_3$	14,60	133,9	0,9203
7	$CH_3\overset{\displaystyle\diagup O}{C}\text{ - }OO\text{ - }\overset{\displaystyle\diagup O}{C}\text{ - }CH_3$	14,65	133,5	0,9014
8	$(CH_3)_2CH\text{-}\overset{\displaystyle\diagup O}{C}\text{ - }OO\,C(CH_3)_3$	14,67	130,5	—
9	$CCl_3\text{-}\overset{\displaystyle\diagup O}{C}\text{ - }OO\,C(CH_3)_3$	14,75	126,8	—
10	$Cl\text{-}\overset{\displaystyle\diagup O}{C}\text{ - }OO\,C(CH_3)_3$	13,70	125,5	0,8876
11	$(CH_3)_3C\,\overset{\displaystyle\diagup O}{C}\text{ - }OO\,C(CH_3)_3$	13,60	121,3	0,8864

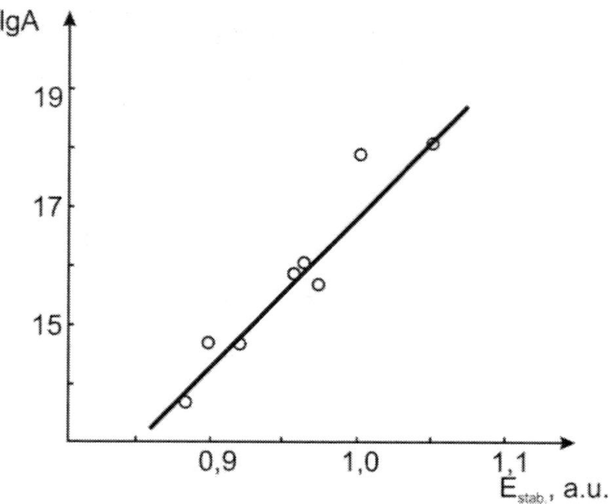

Figure 37. Dependence of *lgA* for peroxides homolysis reaction on their $E_{stab.}$ This graph plotted on the basis of data from Table 30.

$$lgA = \beta E_{stab.} + C \qquad (12)$$

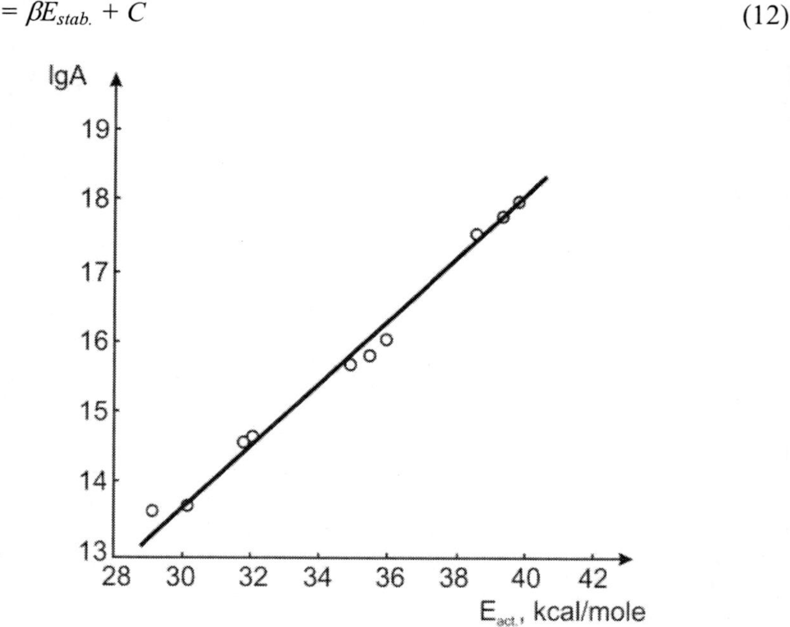

Figure 38. Compensatory ddependence between *lgA* and $E_{act.}$ for the reactions of peroxides homolytical decomposition This graph plotted on the basis of data from Table 30.

Table 31. Dependence of the preexponential factors value for some peroxides homolysis on the stabilization energy of their reactive center

№	Peroxide	lgA	$-E_{stab.}$, a. u.
1	$(CH_3)_3COOC(CH_3)_3$	15,60	1,1426
2	CH_3OOCH_3	15,40	1,0920
3	$C_2H_5OOC_2H_5$	14,18	1,0822
4	$n.C_4H_9OOC_4H_9$	13,98	1,0766

Between lgA and $E_{act.}$ exists the compensatory dependence *(see Figure 38).*

$$lgA = \gamma E_{act.} + D \tag{13}$$

The last equation can be obtained from the eq. eq. (11–12).

Under consideration of peroxides, for which the substitutes have not the sufficient influence on reactive centre, $-C_1\diagdown_{O_\alpha}\diagup^{O_\beta}\diagdown C_2$, the activation entropy of homolysis should greatly depend on weakening the $C_1...O_\beta$ and $C_2...O_\alpha$ non–valency interactions into activated state. In this state the probability of internal rotations around the valency bonds increases and the frequencies of deformative oscillations can greatly change that leads to more friability of the activated complex and its entropy increasing. It seems to us that exactly this moment can explains the influence effect on $\Delta S^{\neq}$ of neighbouring groups under homolytical decompositions of peroxides. Partially this is confirmed by the data of Table 31.

It can be seen from the Table 31, that between the peroxides stabilization energy into equilibrium state and preexponential factors the cymbate dependence exists.

3.11. About Classification of Organic Peroxides

At the present time the peroxy compounds are classified based on only the formal features reflective the nature of substitutes. It's clear, that such classification is conventional. Hardly it's necessary to contradict wholly the importance of this classification since its role as the classifier of knowledge about chemistry of peroxy compounds is evident. However, the formal classification of peroxides can not play the methodological role via the process studies of relation between the reactive ability and their electronic structure.

Below there is classification of the organic peroxides based on electron structure of their reactive center.

Table 32. The values of stabilization energies for some bonds into peroxides

№	Peroxide	Starting state				
		$E_{X\ldots O\alpha}$, a. u.	$E_{X\ldots O\beta}$, a. u.	$E_{C\ldots O\beta}$, a. u.	$E_{O\alpha\ldots O\beta}$, a.u.	$-\Sigma E_1$, a. u.
1	$CH_3C(\!=\!O)\text{-}OO\,C(CH_3)_3$	0,0574	0,0556	0,1460	0,7466	1,0056
2	$CH_2\!=\!CHC(\!=\!O)\text{-}OO\,C(CH_3)_3$	0,0550	0,0448	0,1242	0,7322	0,9562
3	$C_6H_5C(\!=\!O)\text{-}OO\,C(CH_3)_3$	0,0538	0,0532	0,1294	0,7284	0,9658
4	$C_2H_5C(\!=\!O)\text{-}OO\,C(CH_3)_3$	0,0508	0,0458	0,1346	0,7442	0,9854
5	$(CH_3)_3C\,C(\!=\!O)\text{-}OO\,C(CH_3)_3$	0,0566	0,0480	0,1298	0,6524	0,8868
6	$Cl\text{-}C(\!=\!O)\text{-}OO\,C(CH_3)_3$	0,0606	0,0548	0,1294	0,6838	0,9286
7	$H_2C(\text{-}OCH_3)\text{-}OO\,C(CH_3)_3$	0,0564	0,1106	0,1374	0,7464	1,0508
8	$(CH_3)_2NCH_2OOC(CH_3)_3$	0,1140	0,0208	0,1550	0,6405	0,9203
9	$n\text{-}C_4H_9C(\!=\!O)\text{-}OO\,C(CH_3)_3$	0,0572	0,0498	0,1408	0,7626	0,0104
10	$n\text{-}C_3H_7C(\!=\!O)\text{-}OO\,C(CH_3)_3$	0,0590	0,0574	0,1468	0,7656	1,0288

Generally, starting from the results of the conformational analysis of organic peroxides by different structure, these compounds can be divided into two classes. *First class* is the peroxy compounds in which any functional group, group or atom is located into *cis*–position or *cis*–splay position to β–oxygen of the peroxide bridge. In accordance with the conformational imaginations *the second class* is the all rest peroxy compounds for which any functional groups, any atom or group of atoms are not located into *cis*– or *hauch*–position relatively to β–oxygen. Energetic advisability of the *cis*–conformation in comparison with the *trans*–conformation for the peroxides by

general formula $R_1R_2{-}C$ ⋯ ⋯ ⋯ $X{\cdots}O{-}R$, O was explained in *Chapter* 2.

The bond strength of valency–connected and valency–unconnected between themselves atoms into molecule will be characterized by the value of the orbital energy of their paired interaction. The total orbital energy both of valency and non–valency interaction of atoms of the reactive center will be called by the stabilization energy $(E_{stab.})$. This value characterizes the energy advantage of the system under formation of more advantageous conformation.

It can be expected the cymbate dependence between the activation energy of the peroxides homolysis and their stabilization energy, *that is*

$$E_{act.} \approx (E_{IO\alpha-O\beta} + E_{IX\ldots O\beta} + E_{IX\ldots O\alpha} + E_{IC\ldots O\alpha} + E_{IC2\ldots O\alpha})_{act.\ comp.} - (E_{O\alpha-O\beta} + E_{X\ldots O\beta} + E_{X\ldots O\alpha} + E_{IC\ldots O\beta} + E_{C2\ldots O\alpha})_{equil.state} \qquad (14)$$

where E with the respective indexes are orbital energies of bonds of atoms into the molecule which are essential changed via the reaction process. Since into considered peroxides R_1OOR, $R = C(CH_3)_3$ then the value $C_2\ldots O_\alpha$ is approximately constant and can be excluded from the eq. (14). In Table 32 there are values of $E_{stab.}$ for series of peroxides into starting and transition states.

As we can see from this Table 32, the bond $-O-O-$ is greatly weaken into transition state. It is necessary to note, that in homologous series of peroxides $X\ldots O_\alpha$ and $X\ldots O_\beta$ are insignificantly differed upon value; however for different X these interactions are greatly differed upon its value. On *Figure* 39 there is dependence of the activation energy values for homolytic peroxides decomposition on the value $E_{stab.}$ It can be estimated the activation energy of the peroxide homolysis on the basis of value $E_{stab.}$. This procedure requires the calculation both of the equilibrium and activated states. However, taking into account that the potential curves of the peroxides homolytic decomposition ways are not crossed *(see Chapter 3)*, it can be expected the dependence between the activation energy and stabilization energy into equilibrium state.

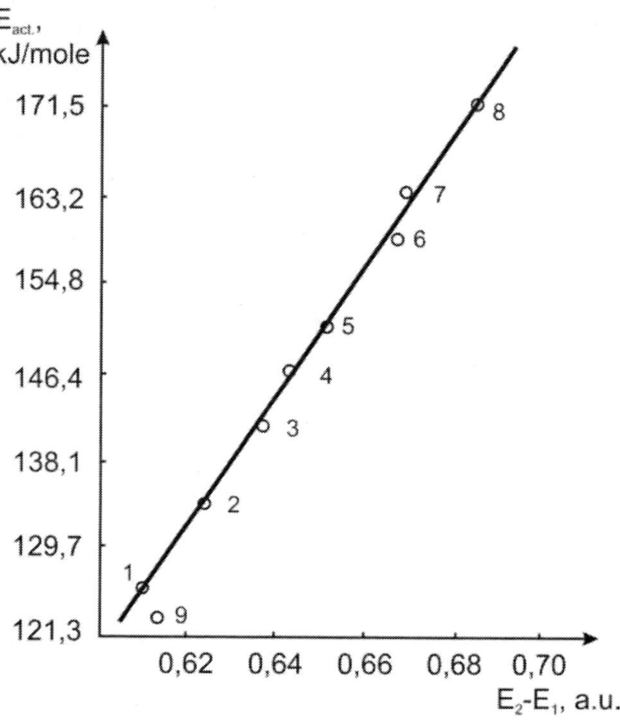

Figure 39. Dependence of $E_{act.}$ *(exp.)* on difference of the stabilization energy into transition and starting states.

1. $(CH_3)_3C-\overset{O}{\overset{||}{C}}-OOC(CH_3)_3$; 2. $(CH_3)_2NCH_2OOC(CH_3)_3$;

3. $CH_2=CH-\overset{O}{\overset{||}{C}}-OOC(CH_3)_3$; 4. $C_2H_5-\overset{O}{\overset{||}{C}}-OOC(CH_3)_3$;

5. $C_6H_5-\overset{O}{\overset{||}{C}}-OOC(CH_3)_3$; 6. $H_3C-\overset{O}{\overset{||}{C}}-OOC(CH_3)_3$;

7. $C_4H_9-\overset{O}{\overset{||}{C}}-OOC(CH_3)_3$; 8. $H_2C\overset{\nearrow OCH_3}{\searrow OO(CH_3)_3}$.

Table 33. The energies values of valency and non–valency bonds for some peroxides into equilibrium state

№	Peroxide	$-E_{O-O}$, a. u.	$\overline{E}_{X\ldots O\alpha}$, a. u.	$\overline{E}_{X\ldots O\beta}$, a. u.	$\overline{E}_{C\ldots O\beta}$, a. u.	$-\sum E$, a. u.
1	$CH_2FOOC(CH_3)_3$	0,8086	0,0590	0,0954	0,1370	1,1000
2	$CH_2ClOOC(CH_3)_3$	0,7212	0,0878	0,2066	0,1510	1,1666
3	$CH_2OCH_3OOC(CH_3)_3$	0,7474	0,0564	0,1106	0,1374	1,0518
4	$CH_2SCH_3OOC(CH_3)_3$	0,7362	0,1752	0,0858	0,1510	1,1482
5	$CH_2NCH_2OOC(CH_3)_3$	0,6405	0,1140	0,0208	0,1550	1,9203
6	$CH_3C(=O)-OO\,C(CH_3)_3$	0,7466	0,0574	0,0556	0,1460	1,0056
7	$C_6H_5C(=O)-OO\,C(CH_3)_3$	0,7284	0,0538	0,0532	0,1294	0,9648
8	$CH_2=CHC(=O)-OO\,C(CH_3)_3$	0,7322	0,0550	0,0448	0,1242	0,9562
9	$CH_2Cl-C(=O)-OOC(CH_3)_3$	0,7408	0,0622	0,0504	0,1370	0,9604
10	$C_2H_5C(=O)-OO\,C(CH_3)_3$	0,7108	0,0508	0,0458	0,1346	0,9754
11	$n\text{-}C_3H_7C(=O)-OO\,C(CH_3)_3$	0,7656	0,0590	0,0574	0,1468	1,0288
12	$n\text{-}C_4H_9C(=O)-OO\,C(CH_3)_3$	0,7622	0,0572	0,0498	0,1408	1,0104
13	$Cl-C(=O)-OO\,C(CH_3)_3$	0,6838	0,0606	0,0548	0,1294	0,9286
14	$(CH_3)_3C\,C(=O)-OO\,C(CH_3)_3$	0,6524	0,0566	0,0480	0,1298	0,8864
15	$H_3C-C(=O)-OOC(CH_3)_3$	0,6494	0,0672	0,0448	0,1460	0,9014
16	$CH_3C(=S)-OOC(CH_3)_3$	0,7498	0,0570	0,0792	0,1614	1,0474
17	$CH_3C(=O)-OO-C(=O)CH_3$	0,6507	0,0563	0,0500	0,1430	0,8900

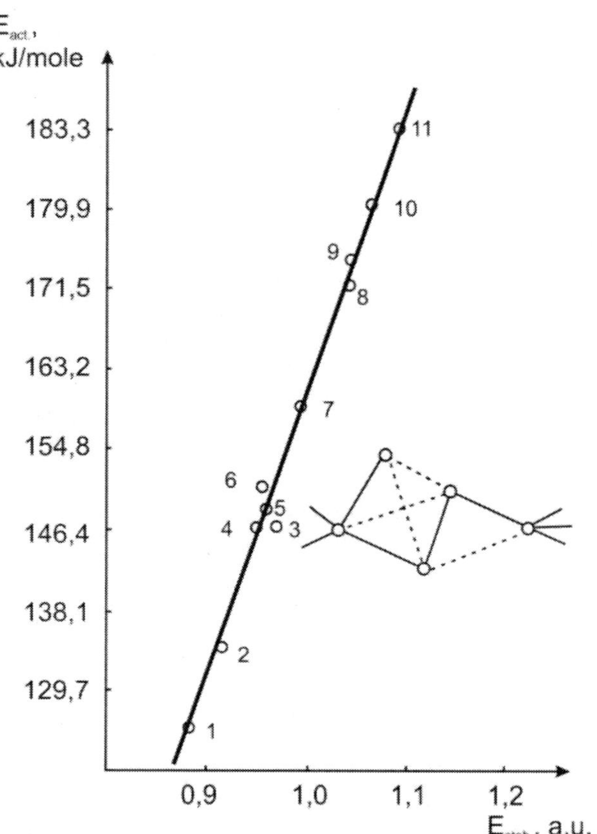

Figure 40. Dependence of the activation energy *(E_act.)* *(experimental)* on stabilization energy *(E_stab.)*.

1. $(CH_3)_3C-\overset{O}{\overset{||}{C}}-OOC(CH_3)_3$; 2. $(CH_3)_2NCH_2OOC(CH_3)_3$;

3. $C_2H_5-\overset{O}{\overset{||}{C}}-OOC(CH_3)_3$; 4. $CH_2=CH-\overset{O}{\overset{||}{C}}-OOC(CH_3)_3$;

5. $CH_2Cl-\overset{O}{\overset{||}{C}}-OOC(CH_3)_3$; 6. $C_6H_5-\overset{O}{\overset{||}{C}}-OOC(CH_3)_3$;

7. $H_3C-\overset{O}{\overset{||}{C}}-OOC(CH_3)_3$; 8. $C_4H_9-\overset{O}{\overset{||}{C}}-OOC(CH_3)_3$; 9. $H_2C\overset{OCH_3}{\underset{OO(CH_3)_3}{}}$;

10. $CH_3C\overset{S}{\underset{OOC(CH_3)_3}{}}$; 11. $CH_2FOOC(CH_3)_3$.

Table 34. The energies values of valency and non–valency bonds for some peroxides into equilibrium state

№	Peroxide	$-E_{C1...O\beta}$, a. u.	$-E_{C2...O\alpha}$, a. u.	$-E_{O\alpha-O\beta}$, a. u.	$-\sum E$, a. u.
1	CH_3OOCH_3	0,2034	0,2034	0,6852	1,0920
2	$C_2H_5OOC_2H_5$	0,1884	0,2036	0,6898	1,0822
3	$C_3H_7OOC_3H_7$	0,2184	0,1888	0,7004	1,0780
4	$(C_2H_5)CH_3CHOOCHCH_3(C_2H_5)$	0,1014	0,1014	0,6976	1,9005
5	$(CH_3)_2CHOOCH(CH_3)_2$	0,1910	0,1910	0,6758	1,0578
6	$C_4H_9OOC_4H_9$	0,1916	0,1960	0,6846	1,0766
7	$\boxed{}\!\!-OO-C(CH_3)_3$	0,1940	0,1794	0,6876	1,0610
8	$(CH_3)_3COOC(CH_3)_3$	0,2293	0,2293	0,6840	1,1426
9	$(CH_3)_3SiOOC(CH_3)_3$	0,2454	0,2704	0,6580	1,1738
10	$(CH_3)_3SiOOSi(CH_3)_3$	0,3491	0,3491	0,6864	1,2846
11	$PhOOPh$	0,1450	0,1448	0,7142	1,0040
12	$PhCH_2OOCH_2Ph$	0,1804	0,2056	0,6966	1,0826
13	$PhCH_2OOC(CH_3)_3$	0,1500	0,1832	0,6236	0,9568
14	H_2O_2	0,0890	0,0890	1,3302	1,5082

In Table 33 there are values of stabilization energies based on some bonds for a series of peroxides.

On Figure 40 there is dependence of activation energy on stabilization energy with taken into account the interaction of atoms of reactive center.

As we can see from Figure 40, a good linear dependence between calculated and experimental energetic parameters permits to determine with sufficient accuracy the activation energy of peroxide homolysis starting from the calculation of the stabilization energy into equilibrium state.

For peroxides by general formula $R1-\underset{R2}{\overset{R3}{C_1}}-OO-\underset{R2}{\overset{R3}{C_2}}-R1$ (where the substituents don't play the essential role) under cyclic reactive center formation the activation energy of homolysis upon $-O-O-$ bond is determined by the strength of valency $-O-O-$ bonds and non–valency bonds $C_1...O_\beta$ and $C_2...O_\beta$. In Table 34 there are values of energies for these bonds. The dependence of experimental activation energies on stabilization energies of peroxides decomposition is represented on Figure 40.

Starting from the calculation of energy stabilization, such dependence permits to estimate the activation energy of homolytic decomposition of a series of organic peroxides. Correlation coefficients of ratios $E_{act.} = f(E_{stab.})$ represented on Figure 40 and Figure 41 are equal to 99,98, *i. e.* the activation energy can be estimated with the error not exceeding the errors at the determination of $E_{act.}$ by experimental way via the corresponding method. However, it is necessary to take into account, that the $E_{act.}$ in many cases is the effective magnitude reflecting the effects influence, medium solvation, «cell» effect, induced decomposition and *at al.* It's clear, that in such cases starting from the stabilization energy the $E_{act..}$ is estimated with greater error. At least, the comparison of $E_{act..}$ values appreciated based on magnitude of $E_{stab.}$ with $E_{act..}$ found via the experimental way, gives the possibility to conclude about like influences on actual $E_{act..}$.

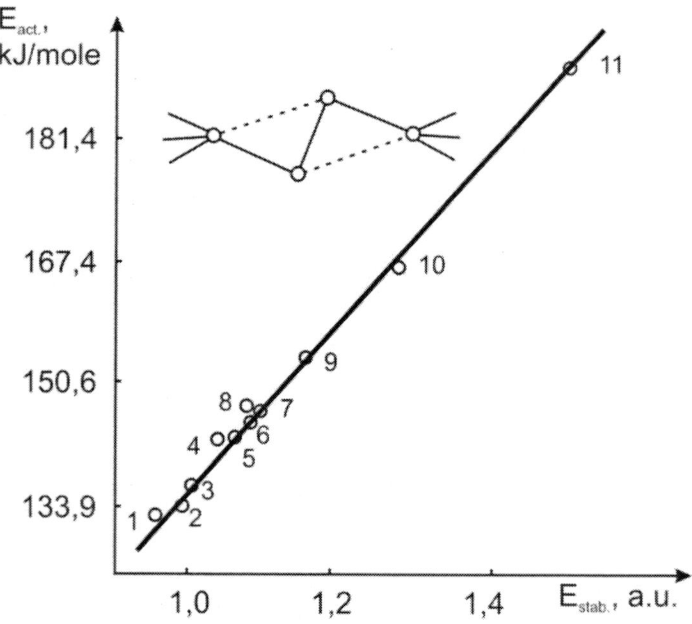

Figure 41. Dependence of the activation energy *(E_act.) (experimental)* on stabilization energy *(E_stab.)* 1. *PhCH_2OOC(CH_3)_3*; 2. *(C_2H_5)CH_3CHOOCHCH_3(C_2H_5)*; 3. *PhOOPh*; 4. *(CH_3)_2CHOOCH(CH_3)_2*; 5. ⬡–*OOC(CH_3)_3*; 6. *PhCH_2OOCH_2Ph*; 7. *CH_3OOCH_3*; 8. *C_2H_5OOC_2H_5*; 9. *(CH_3)_3COOC(CH_3)_3*; 10. *(CH_3)_3SiOOSi(CH_3)_3*; 11. *H_2O_2*.

Table 35. The values of electrostatic energies of reactive centres atoms interaction for peresters into equilibrium and transition states

№	$R-\overset{O}{\underset{1}{C}}-OOC(CH_3)_3$	$E_{1coulomb.}$ of equilibrium state, a. u.				
		$O_\alpha-O_\beta$	$O_\gamma...O_\alpha$	$O_\gamma...O_\beta$	$C_1...O_\beta$	$\sum E_1$
1	CH_3	0,0098	0,0141	0,0102	–0,0146	0,0195
2	C_6H_5	0,0102	0,0145	0,0107	–0,0148	0,0206
3	$(CH_3)_3C$	0,0099	0,0149	0,0107	–0,0146	0,0209
4	$n–C_4H_9$	0,0099	0,0141	0,0104	–0,0145	0,0199
5	$CH_2=CH$	0,0101	0,0144	0,0102	–0,0145	0,0202
6	$n–C_3H_7$	0,0099	0,0138	0,0115	–0,0207	0,0089
7	C_2H_5	0,0098	0,0144	0,0103	–0,0147	0,0196
8	Cl	0,0077	0,0093	0,0078	–0,0148	0,0100
—	——	$E_{2coulomb.}$ of equilibrium state, a. u.				
		$O_\alpha-O_\beta$	$O_\gamma...O_\alpha$	$O_\gamma...O_\beta$	$C_1...O_\beta$	$\sum E_2$
1	CH_3	0,0085	0,0123	0,0131	–0,0170	0,0168
2	C_6H_5	0,0092	0,0137	0,0136	–0,0169	0,0196
3	$(CH_3)_3C$	0,0090	0,0130	0,0132	–0,0170	0,0182
4	$n–C_4H_9$	0,0088	0,0128	0,0133	–0,0163	0,0181
5	$CH_2=CH$	0,0090	0,0130	0,0132	–0,0170	0,0182
6	$n–C_3H_7$	0,0130	0,0130	0,0134	–0,0167	0,0186
7	C_2H_5	0,0087	0,0125	0,0132	–0,0169	–0,0174
8	Cl	0,0070	0,0177	0,0103	–0,0188	0,0062
—	——	$\sum E_2 - \sum E_1$				
1	CH_3	–0,0027				
2	C_6H_5	–0,0010				
3	$(CH_3)_3C$	–0,0027				
4	$n–C_4H_9$	–0,0018				
5	$CH_2=CH$	–0,0020				
6	$n–C_3H_7$	–0,0021				
7	C_2H_5	–0,0021				
8	Cl	–0,0038				

It is necessary to say some words about electrostatic interactions of atoms of reactive centres of investigated peroxides and endowment of this interaction into value of stabilization energy. As we can see from data represented in Table 35, the difference of the electrostatic energies of reactive center into transition and starting states consist of some thousandth of atom unit. This magnitude is incomparably less than the stabilization energy. Probably, it can be whole neglected.

Coulomb energy of atoms interaction only for equilibrium systems of peroxides (see Table *s* 36, 37) is also lesser in comparison with the stabilization energy caused by the quantum–chemical effects (*see* Table 33, 34).

It is necessary to notify that the many used values of decomposition kinetic parameters for investigated peroxides were taken from the reference books [67, 79–83]. In some cases in references there are values for some differing activation energies of peroxides decomposition. This first of all was caused by different experimental conditions (gaseous and liquid phases, kinetic and thermochemical methods of determination of strength of –O–O– bond *at al.*). Under like situations it was taken the averaged value of activation energy. *For example*, for di*tert.*–butylperoxide represented values 163 *kJ/mole*, 156 *kJ/mole*, 157 *kJ/mole*, 158 *kJ/mole*. Normal value consists of 159 *kJ/mole*.

Tables 36. The values of electrostatic interactions for reactive centre of peroxides by structure $R\text{-}\overset{X}{\underset{|}{C}}\text{-}OOC(CH_3)_3$ **into equilibrium state**

№	Peroxide	$E_{coulomb.}$, *a. u.*					ΣE_{colo} *a. u.*
		$O_\alpha\text{–}O_\beta$	$X...O_\alpha$	$X...O_\beta$	$C...O_\beta$	$C\text{–}O_\alpha$	
1	$CH_3\overset{S}{\underset{\parallel}{C}}\text{-}OOC(CH_3)_3$	0,0063	0,0076	0,0086	–0,0077	–0,0062	0,0086
2	$H_2\overset{OCH_3}{\underset{\mid}{C}}\text{-}OOC(CH_3)_3$	0,0106	0,0112	0,0096	–0,0145	–0,0089	0,0082
3	$CH_3\overset{SCH_3}{\underset{\mid}{C}}\text{-}OOC(CH_3)_3$	0,0077	0,0053	0,0059	0,0067	0,0069	0,0053
4	$H_2\overset{N(CH_3)_2}{\underset{\mid}{C}}\text{-}OOC(CH_3)_3$	0,0109	0,0074	0,0071	–0,0127	0,0087	0,0040
5	$H_2\overset{Cl}{\underset{\mid}{C}}\text{-}OOC(CH_3)_3$	0,0075	0,0058	0,0058	–0,0085	–0,0073	0,0033
6	$H_2\overset{F}{\underset{\mid}{C}}\text{-}OOC(CH_3)_3$	0,0098	0,0103	0,0077	–0,0154	–0,0091	0,0033
7	$H_2\overset{H}{\underset{\mid}{C}}\text{-}OOC(CH_3)_3$	0,0095	0,0008	0,0008	–0,0088	–0,0074	–0,0051

Table 37. The values of electrostatic energies of atoms interactions for reactive centre $-XO-OY-$ for peroxides by structure $ROOR_1$

№	Peroxide	$E_{coulomb.}$, a. u.			$\Sigma E_{colomb.}$
		$O_\alpha-O_\beta$	$X...O_\beta$	$Y...O_\alpha$	a. u.
1	CH_3OOCH_3	0,0076	−0,0079	−0,0079	−0,0082
2	$C_2H_5OOC_2H_5$	0,0090	−0,0089	−0,0089	−0,0088
3	$C_3H_7OOC_3H_7$	0,0096	−0,0091	−0,0090	−0,0085
4	$C_4H_9OOC_4H_9$	0,0097	−0,0090	−0,0090	−0,0083
5	$(CH_3)_3COOC(CH_3)_3$	0,0114	−0,0091	−0,0089	−0,0066
6	$(CH_3)_3SiOOC(CH_3)_3$	0,0088	−0,0045	−0,0096	−0,0053
7	$(CH_3)_3SiOOSi(CH_3)_3$	0,0074	−0,0054	−0,0054	−0,0034
8	$PhCH_2OOCH_2Ph$	0,0099	−0,0090	−0,0093	−0,0084
9	$PhCH_2OOPh$	0,0093	−0,0071	−0,0071	−0,0049
10	$OOC(CH_3)_3$ (ecvat.)	0,0102	−0,0081	−0,0083	−0,0102
11	$OOC(CH_3)_3$ (acs.)	0,0073	−0,0097	−0,0076	−0,0100

Since for investigated by us peroxides of *tert.*–alkylperoxycyclopentenes–2 $-OOR$ and epoxy–derived $-OOR$ the stabilization energy is practically the same, their activation energies are also neared, then on Figure 41 only one point for $-OOC(CH_3)_3$ is represented instead of a series of cyclopentenyl peroxides.

The same picture is observed for the peroxide $R_1R_2NCH_2OOR_2$ and for peroxyacetals $R_1CH(OR_2)OOC(CH_3)_3$. Every among these kinds is also represented on Figure 40 by peroxides $(CH_3)_2NCH_2OOC(CH_3)_3$ and $H_2C(OCH_3)OOC(CH_3)_3$ respectively.

Starting from the above–said, the organic peroxides into reactions of homolytical decomposition upon $-O-O-$ bond can be divided into two classes.

First class is the compounds with polycyclic reactive centre

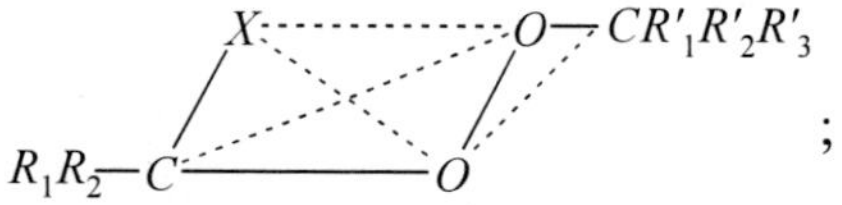

$$X \cdots\cdots O \rightarrow CR'_1R'_2R'_3$$
$$R_1R_2{-}C{-\!\!-\!\!-}O$$

; in other words, this is a case, when the substituents have an essential influence on kinetic parameters of their homolysis via non–valency interactions with the reactive centre.

The second class is the compounds with the bicyclic reactive centre

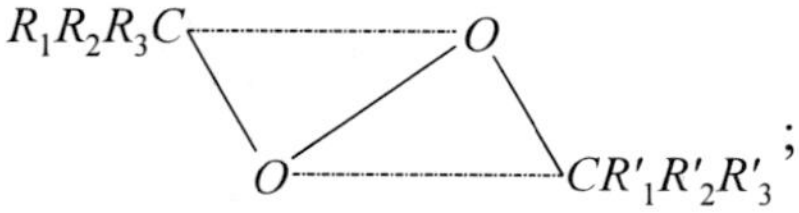

$$R_1R_2R_3C \cdots\cdots O$$
$$O \cdots\cdots CR'_1R'_2R'_3$$

; in other words, this is a case, when the substituents practically haven't an influence on kinetic parameters of these compounds as a result of weak non–valency interactions with the reactive center.

The first indicium of polycyclicity for the reactive center into equilibrium state of peroxides is the conformational advisability of *cis–* and *hauche–* groups of X atoms or atom with respect to peroxide bridge. Secondly, the interaction of X with, for example, O_β–oxygen, that is the formation of «bond», should be accordingly to σ–type of their atom orbitals overlapping. Such classification is advisable not only for the prognostication of homolysis mechanism or purposeful synthesis of peroxides, but, probably is also advisable under the aspect of studying the others reactions of peroxy compounds, since the main principle of such classification is the serious element of the electron structure.

The first class of the peroxy compounds with the polycyclic reactive center includes the peresters, diacylic peroxides, heteroatom–substituted dialkyl(aroil) peroxides, *i. e.* the all peroxy compounds which in β–position to the peroxy bridge have some functional group. The second class includes the peroxides with primary, secondary and tertiary alkyl or aryl substitutes, and also the peroxides with functional groups, located at least not into β–position to the peroxy bridge, *i. e.* their interaction with oxygen atoms of peroxy bridge is insignificantly.

Table 38. Some quantum–chemical characteristics of peroxides with the polycyclic reactive centre

№	Peroxide	$-qO_1$	$-qO_2$	$^*P = K_{AB}$	$-E_R$, a. u.	$-E_K$, a. u.
1	$CH_2ClOOC(CH_3)_3$	0,1672	0,1464	0,9448	0,6690	0,1601
2	$H_2\overset{\diagup SCH_3}{C}-OO\cdot C(CH_3)_3$	0,2118	0,1509	—	—	—
3	$CH_2FOOC(CH_3)_3$	0,2118	0,1509	0,9481	0,6722	0,1607
4	$H_2\overset{\diagup OCH_3}{C}-OO\cdot C(CH_3)_3$	0,2071	0,1665	0,9517	0,6738	0,1613
5	$CH_3\overset{\diagup\diagup S}{C}-OO\cdot C(CH_3)_3$	0,1434	0,1439	0,9496	0,6717	0,1609
6	$C_3H_7\overset{\diagup\diagup O}{C}-OO\cdot C(CH_3)_3$	0,2065	0,1568	0,9542	0,6728	0,1617
7	$C_4H_9\overset{\diagup\diagup O}{C}-OO\cdot C(CH_3)_3$	0,2061	0,1560	0,9537	0,6725	0,1616
8	$CH_3\overset{\diagup\diagup O}{C}-OO\cdot C(CH_3)_3$	0,2068	0,1548	0,9535	0,6724	0,1617
9	$C_2H_5\overset{\diagup\diagup O}{C}-OO\cdot C(CH_3)_3$	0,2058	0,1556	0,9536	0,6724	0,1616
10	$Ph\overset{\diagup\diagup O}{C}-OO\cdot C(CH_3)_3$	0,2086	0,1592	0,9545	0,6725	0,1617
11	$CH_2Cl\overset{\diagup O}{C}-OO\cdot C(CH_3)_3$	0,1971	0,1513	0,9517	0,6722	0,1613
12	$CH_2{=}CH\overset{\diagup\diagup O}{C}-OO\cdot C(CH_3)_3$	0,2125	0,1547	0,9533	0,6714	0,1615
13	$Cl\overset{\diagup\diagup O}{C}-OO\cdot C(CH_3)_3$	0,1766	0,1422	0,9411	0,6681	0,1595
14	$(CH_3)_2NCH_2OOC(CH_3)_3$	0,1984	0,1779	0,9538	0,6745	0,1616
15	$CH_3\overset{\diagup O}{C}-OO\cdot CH_3$	—	—	—	—	—
16	$CH_3\overset{\diagup\diagup O}{C}-OO\cdot\overset{\diagup\diagup O}{C}-CH_3$	0,1620	0,1620	0,9570	0,6749	0,1622
17	$(CH_3)_3C\overset{\diagup\diagup O}{C}-OO-C(CH_3)_3$	0,2089	0,1550	0,9523	0,6717	0,1614
18	$(CH_3)_2CH\overset{\diagup\diagup O}{C}-OO\cdot C(CH_3)_3$	0,2076	0,1569	0,9531	0,6719	0,1615
19	$CCl_3\overset{\diagup\diagup O}{C}-OO\cdot C(CH_3)_3$	—	—	0,9417	—	—
20	$CCl_3OOC(CH_3)_3$	0,1431	0,1436	0,9320	0,6628	0,1579
21	$(CH_3)_2ClCOOC(CH_3)_3$	—	—	0,9417	—	—
22	$CF_3\overset{\diagup\diagup O}{C}-OO\cdot C(CH_3)_3$	0,1751	0,1416	0,9462	0,6708	0,1603

Table 38. (Continued)

№	$-E^v$, a. u.	$+E^I$, a. u.	$+E^N$, a. u.	E_{AB}^{O-O}, a. u.	$-E_{stab.}^{O-O}$, a. u.	$-E_{stab.}$, a. u.	$E_{act.}$, kJ/mole
1	25,0385	12,8464	——	0,2359	0,7212	1,1666	——
2	——	——	——	——	——	——	——
3	25,1383	12,9487	——	0,2372	0,8086	1,1000	——
4	25,1605	12,9718	——	0,2385	0,7474	1,0518	171,5±4,2
5	25,9850	12,7916	——	0,2407	0,7498	1,0474	——
6	25,1395	12,9501	12,7883	0,2386	0,7656	1,0288	——
7	25,1371	12,9476	12,7853	0,2383	0,7622	1,0104	165,7±4,2
8	25,1361	12,9465	12,7853	0,2384	0,7466	1,0056	159,0±4,2
9	25,1357	12,9461	12,7853	0,2383	0,7108	0,9754	146,4±4,2
10	25,1487	12,9595	——	0,2381	0,7284	0,9648	150,6±4,2
11	25,1092	12,9189	——	0,2385	0,7408	0,9604	149,4
12	25,1475	12,9582	——	0,2369	0,7322	0,9562	141,4±4,2
	25,0580	12,8663	——	0,2356	0,6838	0,9286	124,3±4,2
13	25,0490	12,8572	——	0,2341	——	——	——
14	25,1660	12,9775	——	0,2346	0,6405	0,9203	133,9±4,2
15	——	——	——	——	0,6494	0,9014	129,7±4,2
16	25,0596	12,8681	——	0,2433	0,6507	0,8900	129,7±4,2
17	25,1408	12,9513	——	0,2372	0,6524	0,8864	125,5±4,2
18	25,1420	12,9526	——	0,2375	——	——	130,5±4,2
19	——	——	——	——	——	——	——
20	25,9838	12,7903	——	0,2289	——	——	——
21	——	——	——	——	——	——	——
22	25,0448	12,8528	——	0,2378	——	——	——

It is necessary to add, that such classification is previous and needs the checking with the use of any other calculated method. Atom–atomic potential can be used for the search of more energetic advantageous conformers. Under following calculations it should be used only one among other semi–empirical method. Actually, proposed index of reactivity is the resonance component for E_{AB} of molecule under the approximated methods. The value of E_{AB} bond in molecule includes five components: E_{AB}^R is the resonance energy of atoms A and B; E_{AB}^K is their exchange energy; E_{AB}^I is energy of coulomb repulsion of atoms electrons A and B; E_{AB}^v is sum of the potential electrons energy of atom A in a field of atom nuclear B and potential electrons energy of atom B in a field of atom nuclear A; E_{AB}^{N} is energy of atoms A and B atoms nuclears repulsion.

In Table s 38–39 there are values of separate components of E_{AB} for $-O-O-$ bond and proposed by us index – stabilization energy $(E_{stab.})$. The magnitude $E_{stab.}^{-O-O-}$ is actually the resonance energy which characterizes the atoms connecting.

As we can see from the data of Table 40, the experimental activation energies of peroxides thermolysis the best of all correlate with the stabilization energy.

Table 39. Some quantum–chemical characteristics of peroxides with the bicyclic reactive centre

№	Peroxide	$-qO_1$	$-qO_2$	$p^r = K_4$	$-E^R$, a. u.	$-E^K$, a. u.
1	H_2O_2	0,1715	0,1715	1,0052	0,7257	0,1718
2	CH_3OH	0,1740	0,1365	0,9636	0,7041	0,1647
3	$(CH_3)_3SiOOSi(CH_3)_3$	0,1550	0,1509	0,7973	0,6072	0,1351
4	$(CH_3)_3SiOOC(CH_3)_3$	0,2021	0,1550	0,8698	0,6394	0,1474
5	$(CH_3)_3COOC(CH_3)_3$	0,1928	0,1417	0,9570	0,6697	0,1622
6	CH_3OOCH_3	0,1570	0,1928	0,9461	0,6695	0,1603
7	$PhCH_2OOCH_2Ph$	0,1812	0,1570	0,9515	0,6649	0,1612
8	$C_2H_5OOC_2H_5$	0,1717	0,1780	0,9521	0,6683	0,1613
9	$C_3H_7OOC_3H_7$	0,1770	0,1717	0,8512	0,6671	0,1612
10	$C_4H_9OOC_4H_9$	0,1782	0,1770	0,9507	0,6667	0,1611
11	$(CH_3)_2CHOOCH(CH_3)_2$	—	—	—	—	—
12	$\text{—OOC(CH}_3)_3$ (ecv.) (acs.)	0,1769 / 0,2160	0,1886 / 0,1634	0,9515 / 0,9360	0,6667 / 0,6480	0,1612 / 0,1586
13	$PhOOPh$	0,1737	0,1737	0,9462	0,6713	0,1603
14	$PhCH_2OOC(CH_3)_3$	—	—	—	—	—
15	$(C_2H_5CH_3CHO)_2$	—	—	—	—	—
16	$CH_3OOC(CH_3)_3$	0,1739	0,1783	0,9530	0,6724	0,1615

Table 39. Continued

№	$-E^V$, a. u.	$+E^I$, a. u.	$+E^N$, a. u.	E_{AB}^{O-O}, a. u.	$-E_{stab.}^{O-O}$, a. u.	$-E_{stab.}$, a. u.	$E_{act.}$, kJ/mole
1	25,3130	13,0183	12,9153	–0,2769	1,3302	1,5082	205,0±4,2
2	25,2464	12,9497	12,9153	–0,2502	——	——	184,1±4,2
3	25,0312	12,8389	12,7853	–0,1493	0,6824	1,2846	171,5±4,2
4	25,0999	12,9092	——	–0,1922	0,6580	1,1738	163,2±4,2
5	25,1849	12,9971	——	–0,2344	0,6840	1,1426	159,0±4,2
6	25,0393	12,8472	——	–0,2366	0,6852	1,0920	154,8±4,2
7	25,1312	12,9417	——	–0,2303	0,6966	1,0826	——
8	25,0991	12,9087	——	–0,2347	0,6898	1,0826	152,7±4,2
9	25,1206	12,9308	——	–0,2323	0,7004	1,0780	149,4±4,2
10	25,1255	12,9359	——	–0,2321	0,6846	1,0766	143,5±4,2
11	——	——	——	——	0,6758	1,0678	150,6±4,2
	25,1440	12,9549	——	–0,2317	0,6876	1,0610	150,6±4,2
12	25,1723	12,9838	——	–0,2098	0,6876	——	——
13	1072	12,9170	——	–0,2365	0,7142	1,0040	——
14	——	——	——	——	0,6236	0,9568	——
15	——	——	——	——	0,6976	0,9005	142,2±4,2
16	25,1080	12,9179	——	–0,2387	——	——	——

Summarizing the all facts presented in Chapter 3 concerning to the investigation of peroxides reactivity in decomposition homolytical reactions, let us mark the following main moments.

Internuclear distances $-O-O-$ into peroxides in activated complex for reactions of their homolytical decomposition are equal to $1,8 \pm 0,1 \overset{0}{A}$. Forcible arguments in favour of this fact are *ab initio* calculation of cross–sectional of potential energy for the reaction of H_2O_2 way and taking into account the rotational energy of H_2O_2 molecule and also the maximal coming together the energies responsible for $-O-O-$ bonds under internuclear distances $1,8 \overset{0}{A}$.

The stabilization of activated complex in peroxide homolysis reactions is greatly caused by interactions of valency unconnected atoms. Kinetic parameters of the peroxides homolysis can be interpreted from the point of view of taking into account the valency unconnected atoms of the reactive center. Using of such approach gives the possibility to functionally connect the electron structure of peroxides with their reactivity taking into account the space structure.

Proposed classification of organic peroxides is conventional since it is based only on the electron structure of reactive center of peroxides; *however*, it can be useful in practical aspect, *i.e.* it permits to forecast the kinetic parameters of the peroxides homolysis.

Table 40. The values of regression factors for $y = \alpha + \beta x$ and correlation factors *(r)* for peroxides with polycyclic (№№ 1–6) and bicyclic (№№ 7–12) reactive centers

№	y	x	α	β	r
1	$E_{act.}$	$-E_{stab.}$	−139,465	−298,855	0,994
2	$E_{act.}$	$-E_{stab.}^{-O-O-}$	169,649	318,151	0,461
3	$E_{act.}$	$-E_{AB}^{-O-O-}$	697,347	2510,358	0,288
4	$E_{act.}$	$-E_{-O-O-}$	0,000	0,000	0,000
5	$E_{act.}$	$-E_{-O-O-}^{K}$	251,040	836,8	0,078
6	$E_{act.}$	K_{-O-O-}	209,200	−209,200	0,062
7	$E_{act.}$	$-E_{stab.}$	20,514	−122,010	0,999
8	$E_{act.}$	$-E_{stab.}^{-O-O-}$	108,114	−72,408	0,873
9	$E_{act.}$	$-E_{AB}^{-O-O-}$	147,917	−75,015	0,152
10	$E_{act.}$	$-E_{-O-O-}$	26,995	−202,451	0,364
11	$E_{act.}$	$-E_{-O-O-}$	143,206	−141,436	0,086
12	$E_{act.}$	K_{-O-O-}	153,118	10,682	0,037

Note: dimensionality: for №№ 1–12 y – *kJ/mole*; №№ 1–5, №№ 7–11 x – *a. u.*; №№ 6, 12 x – *expansible*; №№ 1–12 α – *kJ/mole*; №№ 1–5, №№ 7–11 β – *kJ/mole·a. u.*; №№ 6, 12 β – *kJ/mole*.

ABOUT REACTIVITY OF SHORT–LIVED OXYRADICALS

In this *Chapter* the questions concerning to the influence of electron structure of oxyradicals on their reactivity in decomposition, isomerization and hydrogen atom break reactions are considered.

4.1. GENERAL INFORMATION ABOUT REACTIVITY OF FREE RADICALS

Main factors determining the stability of radicals are [84]:

- delocalization of unpaired electron;
- steric factors make difficulties for different reactions of radicals;
- deviation from the plain structure decreases the stability of radical.

Heteroradicals in which the unpaired electron localized on atoms differed from the carbon are like to the alkyl ones upon behaviour and their reactions can be classified similarly.

4.2. EXPERIMENTAL INFORMATION ABOUT OXYRADICALS

The methods of oxyradicals identification usually are indirect. The direct mass–spectrometric determination of RO was described only for the phenoxy–radical forming under the anisole pyrolysis, and for the methoxy–radical obtained under photosensitized by *Mercury* decomposition of methanol and dimethyl ether [74]. Formed under thermo– and photodecomposition RO depending on its structure are undergoing to different reactions (decomposition, hydrogen break, addition to olefins, disproportionation and isomerization).

Decomposition of RO is possible in two directions:

- with the detachment of hydrogen atom $RCH_2O \rightarrow H + RCHO$;
- with the detachment of alkyl radical $RCH_2O \rightarrow R^{\cdot} + CH_2O$.

On the basis of data obtained under decomposition of series non–symmetric alkoxy–*tert.*–butylperoxides into solution of cyclohexene at 468 ^0K characterizing the ratio of reactions products of hydrogen atom break and decomposition of forming oxyradicals it was determined the order of stability for simple alkoxy–radicals [74]: $CH_3O > C_2H_5O > n–C_4H_9O > iso–C_3H_7O > iso–C_4H_9O > tert.–C_4H_9O$. Experimentally obtained order on stability coincides with the thermo–chemical data of oxyradicals, in other words, rather decomposing radicals have less decomposition enthalpy. In liquid phase the decomposition reactions are less typical for RO (methoxy–, ethoxy–, n– and *iso*–propoxy– and also n–butoxy–) radicals. These radicals generated in different solvents do not undergo to the decomposition even at 393–428 K [74]. In the presence of reagents with easy breaking $C–H$ bonds it is possible the break of hydrogen by RO radicals with the alcohol formation. Into gaseous phase the hydrogen break reactions a great extent depend on RO stability. Methoxyradicals are the most stable.

4.3. ESTIMATION OF FREE RADICALS REACTIVITY

In order to estimate the reactivity of molecules in different reactions often their quantum–chemical indexes (charges, polarizabilities) are used. General approach to the determination of these indexes was formulated in *Coulsone's* and *Longuet–Higgins's* works [85–86]. The all indexes characterize practically isolated molecule and their role is reduced to the qualitative correlation of experimental data for the limited reactive series [87]. Nevertheless, the indexes of molecules reactivity are used by authors under interpretation of the experimental material on chemical reactions. Number of works concerning to the application of quantum–chemical indexes for the interpretation of reactivity of σ–radicals into substitution or addition reactions and *ect.* in references is insignificant. There are only several works in this aspect [88].

Let us consider the alkoxyl radicals under this aspect in some detail. Data of Table 41 show, that between the reactivity of alkoxyl radicals into reaction of hydrogen atom break and the charge on oxygen with the unshared electron it is observed the antibate dependence. The charge increasing on the oxygen of *RO* radical decreases its reactivity.

To understand the sense of this ratio is possible only in one case, when to lay that $\Delta E \approx q_r \Delta \alpha_r$ (here ΔE is the disturbance energy of addition or hydrogen atom break from the molecule for like series of radicals with one molecule; q_r are charges of radicals atoms with the unshared electron; $\Delta \alpha_r$ the changes of coulomb's integral for presented series of reactions). This means, that the *RO* radical is polar particle, $\Delta \alpha_r = const$ for this series of reaction. From the other hand, perhaps the localization of charge on the atom of oxygen permits to consider the *RO* as «*dotty*» reagent, which can be modelled by one *AO* [87].

Table 41. Dependence of the comparative reactivity of *RO* radicals into hydrogen atom break and isomerization reactions on the magnitude of charge on oxygen atom and the values of lower free orbital (*lfo*) energies

№	Radical	RO+cyclohexane, % of ROH yield; T = 468 K (gas. phase) [74]	$-q_0$	$\varepsilon_{(lfo)}$, a. u.	$-\Delta H^{\neq}$ of isomerization, kJ/mole	Forming radical under isomerization
1	CH_3O	76	0,1714	0,5266	31,4	CH_2OH
2	C_2H_5O	65	0,1957	0,5112	39,7	$CH(CH_3)OH$
3	$n–C_3H_7O$	—	0,2011	0,5093	—	—
4	$n–C_4H_9O$	30	0,2027	0,5078	—	—
5	$iso–C_3H_7O$	19	0,2130	0,4936	50,2	$CH(CH_3)_2OH$
6	$iso–C_4H_9O$	6	0,2162	0,4907	—	—
7	$(CH_3)_3CO$	—	0,2267	—	—	—

It is observed some parallel between decreasing of the $RO^.$ radical's reactivity into reactions of hydrogen atom break and decreasing of their lower free orbital energy *(see* Table 41*)*. Probably, this moment can be explained by enough great endowment of this level via the interaction reaction of $RO^.$ with *RH*.

Thus, if to assume, that the charge on oxygen atom can be used as the index of the $RO^.$ radicals reactivity, then the oxyacyl radicals $RC{\overset{O}{\underset{O^.}{\lessgtr}}}$, for which the charge on oxygen atoms is in a ranges $(0,30–0,39)^*$, should take place after the alkoxyl radicals upon its activity depending on nature of *R*.

Unfortunately, we have not any experimental data concerning to the hydrogen atom break by these radicals.

For the radicals by general structure $R_1H(R_2O)CO^.$ the charge on oxygen having the unshared electron depending on the substitutes is in the range of $-0,20–0,22$; in other words, accordingly to our assumptions these particles upon their reactivity are on a level of radicals $n–C_3H_7O$ and $(CH_3)_3CO^.$. However, as our investigations showed, the radicals by $R_1H(R_2O)CO^.$ type generally can not break the hydrogen atom even from the compounds with the low strength of *C–H* bond. The same phenomenon is observed also in a case of radical $(CH_3)_2NCH_2O^.$. The charge on oxygen is equal to $-0,2149$, that is accordingly to its magnitude the reactivity of this radical should be enough high. However, this is not correct. The radicals of this type can not able break even weakly connected atom of hydrogen. These experimental facts prejudice the use of oxygen atom charge with unshared electron as index of relative reactivity of radicals even within one type of radicals range.

This implyies that, possibly under determined conditions, in other words, for some narrow series of radicals with near upon its nature substitutes, the charge on atoms with unshared electron can be used as the index of reactivity. However, this index as more universal even within one series of radicals with different substitutes becomes by ineligible for the reactions with the molecule.

Table 42. The values of spin densities on atoms having the unshared electron for some radicals

Radical	Atom with the unshared electron	Full spin density	Spin density on P–*AO*
OH	*O*	1,0476	1,0000
CH₃	*C*	1,1299	1,0000
CH₃O	*O*	0,9616	0,9289

4.4. ABOUT ENERGY OF MO FOR UNSHARED ELECTRON IN RADICALS

The energy of *MO* for unshared electron a great extent depends on nature of the substitute *R.*. For a example, at substitutes donor properties increasing in a range *HO* > *CH₃O* > *C₂H₅O* > *(CH₃)₃CO* the energy of *MO* of free electron is increased in order –0,7258 < –0,5929 < –0,5617 < –0,4919 *(a. u.)*.

The analysis of spin density distribution in radicals *(see* Table 42*)* shows, that in some radicals, for example, *OH* and *CH₃*, the electron wholly localized on one among atomic obitals since the spin density is equal to 1.

In radicals with more complicated substitutes the unshared electron to some extent delocalized in such manner that the spin density of unshared electron localizing on corresponding *AO* is less than 1 *(see* Table 43*)*.

The same conclusion can be done starting from the analysis of proper vectors and proper values of *MO* of radicals. Thus, it is follow from the data of Table 44 that the unshared electron in radicals *OH* and *CH₃* localized on one *MO*, since the coefficients are equal to 1. For more complicated radicals, the *p–AO* on which there is an unshared electron, is part of the series of molecular orbitals *(see* Table 44*)*.

Probably, that under consideration of radicals reactivity into reactions it is necessary to take into account the endowments of *AO* of unshared electron and also the other *MO*. The energy of effective *MO* for unshared electron can be represented by the expression:

$$E_{eff.} = c_{i1}^2 E_1 + c_{i2}^2 E_2 + ... + c_{ig}^2 E_g \tag{15}$$

where c_{ig} are proper vectors i *AO* in g *MO*; E_g is energy g *MO*.

Obtained values of $E_{eff.}$ for the free radicals by different nature are represented in Table 45. As we can see from the Table 45, the magnitudes of $E_{eff.}$ are differed upon their values for the radicals and, probably, in connection with this fact, likely should be differed upon its reactivities via some chemical transformations. For example, it is seen from the data of Table 46, that the cymbate dependence is observed between the $E_{eff.}$ and the activation energy for the hydrogen break reactions whereas none regularity is not observed between the higher occupied orbital *(hoo)*, lower free orbital *(lfo)*, free electron orbital energy and activation energy.

         A. A. Turovsky, L. I. Bazylyak, A.R. Kytsya et al.

Table 43. The spin density distribution on atoms of radicals and on atomic orbitals of oxygen atom

Atom	R/AO	$\overset{\bullet}{O}$–H	$CH_3\overset{\bullet}{O}$	$(CH_3O)CH_2\overset{\bullet}{O}$
$\overset{\bullet}{O}$	—	1,0476	0,9635	0,9711
$\overset{\bullet}{C}$	—	—	–0,1047	–0,1084
H	—	–0,0476	0,0944	0,0728
H	—	—	0,0234	0,0728
O^*	—	—	—	–0,0072
C^*	—	—	—	–0,0011
H	—	—	0,0234	–0,0005
H	—	—	—	0,0002
H	—	—	—	0,0002
$\overset{\bullet}{O}$	S	0,0127	0,0172	0,0167
	P_x	0,0000	0,0013	0,0035
	P_y	0,0350	0,0142	0,0151
	P_z	1,0000	0,9308	0,9358

The analysis of sufficiently great number of references on kinetics of bimolecular reactions of hydrogen break from the molecules brings us to the conclusion that the series of the radicals' activity represented in Table 46 is kept in the best way in a case of reactions with paraffins, olefins, cyclic saturated hydrocarbons. The effective energy of *MO* for unshared electron can be used as the index of reactivity in reactions of atom break by radical only for series of radicals with one molecule. In these cases it is supposed that the same interaction effects take place into activated complex of the reaction.

Let us consider a question of the short–lived radicals' reactivity under some other aspect.

The disturbance energy for the synchronous exchange interaction of two particles with the reactive centres s and t is expressed by the equation [43]:

Table 44. The values of some calculated parameters of free radicals

Radical	Atom with unshared electron	Type of AO	Spin		1	2	3	4
$\overset{\bullet}{H}O$	$\overset{\bullet}{O}$	P	α	$E_{a.e.}^{MO}$	−1,4481	−0,7449	−0,7258	−0,6618
				C_{AO}	0,0000	0,0000	1,0000	0,0000
			β	$E_{a.e.}^{MO}$	−1,3174	−0,7076	−0,6269	0,1469
				C_{AO}	0,0000	0,0000	0,0000	1,0000
$\overset{\bullet}{C}H_3$	$\overset{\bullet}{C}$	P	α	$E_{a.e.}^{MO}$	−1,1453	−0,7345	−0,7327	−0,4826
				C_{AO}	0,0000	0,0000	1,0000	−1,0000
			β	$E_{a.e.}^{MO}$	−1,3119	−0,7266	−0,7247	0,1421
				C_{AO}	0,0000	0,0000	0,0000	−1,0000
$CH_3\overset{\bullet}{O}$	$\overset{\bullet}{O}$	P	α	$E_{a.e.}^{MO}$	−1,6311	−1,1101	−0,8695	−0,8449
				C_{AO}	−0,0003	0,0062	0,6594	0,0000
			β	$E_{a.e.}^{MO}$	−1,5500	−1,0681	−0,8335	−0,7697
				C_{AO}	0,0003	0,0009	0,0000	0,2304

Table 44. Continued

Radi-cal	Atom with unshared electron	Type of AO	Spin		5	6	7	8	9	10	11
$\overset{\bullet}{H}O$	$\overset{\bullet}{O}$	P	α	$E_{a.e.}^{MO}$	0,2380						
				C_{AO}	0,0000						
			β	$E_{a.e.}^{MO}$	0,2586						
				C_{AO}	0,0000						
$\overset{\bullet}{C}H_3$	$\overset{\bullet}{C}$	P	α	$E_{a.e.}^{MO}$	0,2372	0,3500	0,3520				
				C_{AO}	0,0000	0,0000	0,0000				
			β	$E_{a.e.}^{MO}$	0,2740	0,3615	0,3626				
				C_{AO}	0,0000	0,0000	0,0000				
$CH_3\overset{\bullet}{O}$	$\overset{\bullet}{O}$	P	α	$E_{a.e.}^{MO}$	−0,7050	−0,5909	−0,5510	0,2370	0,2979	0,3043	0,3561
				C_{AO}	−0,0202	0,7394	0,0000	−0,0094	−0,1340	0,0000	0,0000
			β	$E_{a.e.}^{MO}$	−0,6739	−0,5266	0,1126	0,2455	0,3065	0,3398	0,3768
				C_{AO}	−0,0012	0,0000	−0,8568	−0,0225	0,0000	−0,4605	0,0141

Table 45. The values of $E_{eff.}$ and E_g for free radicals

№	Radical	$-E_{f.\,e.}$, a. u.	$+E$, a. u.	$[E_{eff.}]$, a. u.
1	OH	0,7258	0	0,7258
2	CH_3O	0,7012	0,0053	0,7066
3	C_2H_5O	0,6964	0,0039	0,7003
4	$n–C_4H_9O$	0,6905	0,0028	0,6933
5	$n–C_3H_7O$	0,6852	0,0039	0,6891
6	$tert–C_4H_9O$	0,6844	0,0032	0,6876
7	$Cl-C{\overset{O}{\underset{O\cdot}{\lessgtr}}}$	0,6354	0,0634	0,6666
8	$H_2C{=}CH-C{\overset{O}{\underset{O\cdot}{\lessgtr}}}$	0,5801	0,0700	0,6501
9	$tert.-C_4H_9-C{\overset{O}{\underset{O\cdot}{\lessgtr}}}$	0,5779	0,0481	0,6260
10	$H_3C-C{\overset{O}{\underset{O\cdot}{\lessgtr}}}$	0,5806	0,0748	0,6554
11	$(CH_3)_3SiO$	0,5917	0,0090	0,6007
12	C_6H_5	0,4500	0,1015	0,5515
13	CH_3	0,4826	0	0,4826
14	$n–C_6H_{13}$	0,4241	0,0529	0,4770

Table 46. The values of some energies for MO and $E_{eff.}$ for radicals and activation energy for reaction $(R, RO) + CH_4 \rightarrow (RH, ROH) + CH_3$

№	Radical	$-E_{hoo}$, a. u.	$-E_{lfo}$, a. u.	$-E_{f.\,e.}$, a. u.	$-E_{eff.}$, a. u.	$E_{act.}$, kJ/mole
1	HO	0,6618	0,1463	0,7258	0,7258	25,9
2	CH_3O	0,5518	0,1176	0,5929	0,7066	49,4
3	C_2H_5O	0,5320	0,1351	0,5617	0,7003	—
4	$(CH_3)_3CO$	0,4919	0,1538	0,4919	0,6876	—
5	$CH_3C{\overset{O}{\underset{O\cdot}{\lessgtr}}}$	0,5355	0,0300	0,5674	0,6554	—
6	$(CH_3)_3CC{\overset{O}{\underset{O\cdot}{\lessgtr}}}$	0,4938	−0,0198	0,6485	0,6260	—
7	C_6H_5	0,4033	0,0882	0,4033	0,5515	51,9
8	CH_3	0,4826	0,1421	0,4826	0,4826	61,1
9[*]	$n–C_6H_{13}$	0,4258	0,1956	0,4258	0,4770	71,1

[*] It was estimated accordingly to additivity method.

$$\Delta E = -\frac{q_s q_t}{R_{st}\varepsilon} - \frac{q_{s'} q_{t'}}{R_{s't'}\varepsilon} + \sum_m \sum_n \frac{v_{mn}\left(c_s^m c_t^n \Delta\beta_{st} + c_{s'}^m c_{t'}^n \Delta\beta_{s't'}\right)^2}{E_m^* - E_n^*} \qquad (16)$$

where q are the charges of atoms; R is the distances between atoms; c are the coefficients of atomic orbitals of atoms in different MO of corresponding molecules; $\Delta\beta$ is the change of the resonance integral at interaction of atom's orbitals; E^* is the magnitude characterizing the energies of different molecular orbitals m and n in isolated molecules in the same medium which is used for the reaction carrying out. The value $v = 0$, if the sum of the number of electrons in m and n is equal to 0 or 4; $v = 1$ if the sum of electrons is equal to 1 or 3 and $v = 2$, if the sum of electrons is equal to 2. The last excludes the consideration of interaction of free orbitals of reagents with twice occupied orbitals [43].

In a case of the radicals reaction the calculation should take into account the interactions of one–occupied orbital both with occupied and vacant orbitals, since such interactions lead to the change of energy. In a case of the same separating of energetic levels between the orbital with one electron and vacant or twice occupied orbital these interactions are the same. If the separating of levels is different, then the vicinal orbital interaction will be dominating [43].

Usually the most endowment into the change of covalent energetic term inserts the specific interaction of any determined pair of MO. These orbitals are called «*stereodetermining*». For the molecules with the closed shells the boundary orbitals, *HEMO* and *LFMO* are «stereodetermining», but for the radicals is not always like this.

Let us consider the specific examples of the radical's reactions. Under reaction of OH radical with CH_4 in accordance with the HMO the following interactions are taken into account:

$$E_1 = \frac{(-0{,}3671)^2}{-0{,}7258 - 0{,}2904} + \frac{(0{,}5179)^2}{-0{,}7258 - 0{,}3374} + \frac{(0{,}3040)^2}{-0{,}7258 - 0{,}3374} + \frac{(0{,}0019)^2}{-0{,}7258 - 0{,}3374} = 0{,}2433 a.u.$$

(Free electron and *LFMO* of CH_4);

$$E_2 = \frac{(0{,}3395)^2}{0{,}7258 - 1{,}2915} + \frac{(0{,}0002)^2}{0{,}7258 - 0{,}7193} + \frac{(0{,}3994)^2}{0{,}7258 - 0{,}7193} + \frac{(0{,}4795)^2}{0{,}7258 - 0{,}7193} = -30{,}7865 a.u.$$

(Free electron and *HEMO* of CH_4);

$$E_3 = \frac{(0,3395)^2}{-1,2915-0,1463} + \frac{(0,0002)^2}{-0,7193-0,1463} + \frac{(0,4795)^2}{-0,7193-0,1463} + \frac{(0,3994)^2}{-0,7193-0,1463} = 0,5466 a.u.$$

(HEMO of CH$_4$ and LFMO of radical OH).

It can be seen that the most endowment into the value of the energy disturbance of the particles interaction brings the term E_2. The results of calculations of other radicals and atoms with the methane are represented in Table 47.

Integrals of the intermolecular interaction overlapping have been calculated accordingly to *Batsanov* [89] and the resonance integrals in accordance with the *Popl* [90]. The distance between the reacting particles was taken equal to the sum of *Van der Waals* radiuses of final reacting atoms of these particles.

It can be seen from Table 47, that some less endowment into the orbital energetic term of interaction has the levels' interaction of free electron of radical and the *LFMO* of methane. It is interesting, that this term it is not enough changed depending on nature of considered radicals and atoms. The levels' interactions of free electrons and *HEMO* of methane have the ponderable endowment into the orbital energy.

The data of Table 47 permit to conclude that the decrease of the disturbance energy of the interacting particles is symbate to increasing of their experimental activation energies.

Probably, the transition state of these reactions is similar to the starting reagents, in other words in this state the bond *C–H* into methane is little loosened, and the bond *R–H* is little similar to the bond into the final products.

Table 47. The dependence of activation energy for reaction $R + CH_4 \rightarrow$ $RH + CH_3$ on disturbance energy of system

№	R	E_1, a. u.	E_2, a. u.	E_3, a. u.	$E_{distib.}$, a. u.	$E_{act.,}$ kJ/mole (gas. phase) [67]
1	CH_3O	0,2459	0,3835	0,5124	1,1418	46,0
2	C_6H_5	0,5660	1,4857	1,0076	3,0593	51,9
3	H	0,7822	1,8176	0,5734	3,1732	33,5–50,2
4	CH_3	0,6938	1,4387	1,1958	3,3283	61,1

It is necessary to notify, that the disturbance energy of the considered reagents is greatly determined by the energetic level of free electron. In a series of different radicals reactions with one molecule the relative reactive ability will be greatly determined by the energy of the level of free electron of radical.

For the case of radical reaction with a molecule, when the equipped level of free radical reacts with the *HEMO* of molecule and under condition that these levels practically localized, the equation of the disturbance of molecular orbitals *(DMO)* will be as follow:

$$\Delta E_{disturb.} = \frac{c_s^2 c_t^2 \Delta \beta_{st}^2}{E_s - E_t} \qquad (17)$$

in other words, if it's considered the reaction for a range of radicals with the same molecule, then E_t and c_t^2 are the constants and

$$\Delta E_{disturb.} = \frac{c_{s'}^i \Delta \beta_{st}}{E_{s'}}.$$

For the same type of reactions it can be laid that $\Delta \beta_{st} = const$, then

$$E_{disturb.} = \frac{c^2}{E_s''} \qquad (18)$$

If the *AO* of free electron is strongly delocalized upon others *MO* then

$$\Delta E_{disturb.} = \Sigma \frac{c'^2}{E_s''} \qquad (19)$$

Proposed index of the reactivity is as follows $E = \sum_g c_{ig}^2 E_g$.

It is interesting that the disturbance energy for reaction of a series atoms and radicals with the polar molecule (H_2O) is also symbate in a range of their exponential values of activation energy (see Table 48).

Table 48. Dependence of activation energy of reactions on their disturbance energy

Reaction	$\Delta E_{disturb.}$, a. u.	$E_{act.}$, kJ/mole (gaseous phase), [67]
$F + H_2O \rightarrow HF + OH$	1,2728	25,1; 29,3
$O + H_2O \rightarrow OH + OH$	1,8739	77,4 ± 6,3; 87,19; 74,27; 75,7 ± 4,2
$H + H_2O \rightarrow H_2 + OH$	2,8318	90,4 ± 6,3; 90,58
$CH_3 + H_2O \rightarrow CH_4 + OH$	2,9311	94,39

4.5. ABOUT RATIO BETWEEN THE ORBITAL ENERGY AND THE ENERGY OF ELECTRONS REPULSION FOR RADICALS AND ATOMS

Full electronic energy E for particles with open shell under approximation of MO is represented by the expression [69]:

$$E = \sum_{\mu} \varepsilon_\mu + \sum_{\mu < \nu} \sum I_{\mu\nu} - \overset{\uparrow\uparrow}{\sum_{\mu < \nu} \sum} K_{\mu\nu} \tag{20}$$

Under this approximation the coefficients at AO are selected so as to minimize the full electronic energy E. The summarizing is carried out upon the all occupied MO, and the last sum is taken upon the all MO with the same spin.

It is completely logically to suppose that the relative stability of radicals and free atoms depends on the ratio of full orbital energy and the electrons repulsion energy. The more this ratio the more stable is radical and the less its reactivity in any transition reactions.

Since the ε_μ is full energy of electron, lying in orbital φ_μ, caused by its movement (kinetic energy), by gravitation to all cores and by the repulsion from all electrons, and the second term of equation (20) characterizes the average interelectron repulsion of electrons of whole system, then it can be expected that the stability will be depend on the ratio of these two energetic terms.

Table 49. The total orbital energy ($\sum \varepsilon_\mu$), average repulsion energy of all electrons and their ratio (γ)

№	Radical or atom	$-\sum \varepsilon_\mu$, a. u.	$\sum\sum\limits_{\mu<\nu}\left(I_{\mu\nu}-K_{\mu\nu}\right)$, a. u.	γ	$E_{act.}$ [67], kJ/mole; $R + CH_4 \rightarrow RH + CH_3$
1	F	6,3748	21,1743	0,3011	5,06 ± 0,33
2	O	4,6961	13,3859	0,3508	27,6 ± 6,3
3	OH	6,2325	11,9300	0,5224	20,9; 21,3 ± 1,7
4	$(CH_3)_3CC{\overset{O}{\underset{O\bullet}{\lessgtr}}}$	3,8139	6,8593	0,5560	—
5	$CH_3C{\overset{O}{\underset{O\bullet}{\lessgtr}}}$	21,2283	29,7116	0,7145	—
6	CH_3O	11,7243	14,8862	0,7876	46,0
7	C_6H_5	26,1656	18,4432	1,4187	51,9
8	CH_3	5,6655	3,2102	1,7648	61,1

It can be seen from the data of Table 49, that the more this ratio the less is the reactivity of radical or atom into reactions of hydrogen breaking from the molecule of methane. If in a range of the radicals reactions with any molecule different other effects of interaction (electrostatic and space) can be approximately the same or insignificant, then this index γ can play the definite role under estimation of relative reactivity of radicals and atoms. For example, from the data of Table 50 it is seen, that the range of the radicals and atoms reactivity under interaction with sufficiently polar molecule HCl is practically not dislocated; in other words it is the same as into the reactions with the alkanes. It is follows from this that the endowments of the polar effects into $E_{act.}$ for the presented series of reactions are the same.

Table 50. The value of γ and activation energies for the reactions
$$R + HCl \rightarrow RH + Cl$$

№	R	γ	$E_{act.}$ [67], kJ/mole
1	F	0,3011	2,5
2	O	0,3508	3,8
3	OH	0,5224	5,4
4	CH_3	1,4187	18,4; 20,1; 20,5; 20,9

4.6. ABOUT STABILIZATION ENERGY OF FREE RADICALS

Reactivity of free radicals in any substitution or addition reactions depends on not only the energy of level of free electron, but also is determined by other levels of atom energy having the unpaired electron. In a range of free radicals reactions with any molecule the relative reactive ability of radicals can be determined by nature of themselves free radicals. Reactivity of radical can be characterized by energy depending on the interaction degree of AO of atom having the unpaired electron with the AO of neighbouring atoms valence free from it. Under concerned sense this reaction will be called as the stabilization energy of free radical.

Under consideration the series of alkoxyl radicals as the criterion of reactivity let's accept the stabilization energy, which is determined by total energy of the interaction of fourth AO of oxygen atom having the unpaired electron with the AO of other atoms valence free from it. It is clear, that in every concrete case it is necessary to take into account the most essential interactions which are easy discovered as a result of analysis the results of free radical calculation.

Let's consider the concrete examples of radicals which are the primary products of the homolysis of investigated peroxides. For the radical the most essential are the non–valence interactions of atoms $H_2...O_1$ (–0,2154 $a.$ $u.$); $O_1...O_5$ (–0,1107 $a.$ $u.$); $C_6...O_1$ (–0,1404 $a.$ $u.$); $H_2...O_5$ (–0,0384 $a.$ $u.$). In generally, this consists of –0,5049 $a.$ $u.$

Table 51. The stabilization energies for the radicals RO

№	Radical	$-E_{stab.}$ upon non–valence bonds $O...H$, $a.$ $u.$	$-E_{stab.}$ upon $C–H$ bonds, $a.$ $u.$	ΔH, enthalpy of hydrogen breaking, $kJ/mole$
1	CH_3O	0,0031	0,3942	104,6
2	$CH_3–CH_2O$	0,0035	0,3849	87,9
3	$CH_3–CH_2–CH_2O$	0,0038	0,3664	79,5
4	$CH_3–(CH_2)_2–CH_2O$	0,0040	0,3884	96,2

For the radical $\begin{array}{c} H \quad\quad N(CH_3)_2 \\ \diagdown \quad \diagup_{3} \\ C \\ \diagup_1 \diagdown \\ H_6 \quad\quad O_2{}^{\bullet} \end{array}$ the most endowment into stabilization

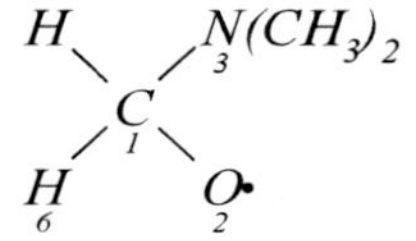

energy give the non–valence interactions $C_3...O_2$ (–0,1493 $a.\ u.$); $C_4...O_2$ (0,0473 $a.\ u.$); $C_5...O_2$ (0,0647 $a.\ u.$); $H_6...O_2$ (0,2038 $a.\ u.$). In generally, this consists of –0,4551 $a.\ u.$

For radicals by $R-C \begin{smallmatrix} \cdot\cdot O_1 \\ O_2 \end{smallmatrix}$ type the stabilization energy is represented in

Table 52 (it was taken into account the most essential interactions $O_1...O_2$). As we can see from Table 52, the stabilization energy for this type of radical is within the range of 0,14–0,18 $a.\ u.$ The activity of these radicals into reactions of the atom breaking is increased in turn of numbers rise in Table 52.

For the radicals RO the stabilization energies are represented in Table 51. Reactivity of RO is greatly caused by the interaction degree of AO of one among atoms of substitute with the AO of oxygen atom having the unpaired electron.

As we can see from Table 49, AO of hydrogen atom is weakly overlapped with s and p of AO of oxygen atom; that is why the stabilization energy of these radicals is insignificant and is within the interval 0,004–0,005 $a.\ u.$; this is much lower than in a case considered some above. In connection with this fact such type of free radicals is extraordinarily reactive.

In a case of radical $(CH_3)_3CO$ the stabilization energy is some higher (–0,1011 $a.\ u.$), that is upon its reactivity into reactions of atom breaking it less reactive than the considered radicals RO.

For the radical $(CH_3)_3SiO$ the value of stabilization energy is very insignificant (–0,0111 $a.\ u.$). This means, that the radical is low selective and very reactive into reactions of hydrogen atom breaking. This is confirmed by our experimental investigations on kinetics and mechanism of homolytical decomposition of silicon–containing peroxides *(see next* Chapter 5).

Under considered sense the stabilization energy for OH radical is wholly equal to zero, that is the presented radical is the most reactive among the all alkoxyl free radicals.

For radical CH_3OO the most endowment into stabilization energy gives the interaction $C...O_2$ (–0,2384 $a.\ u.$); that is, this radical upon its reactivity in

reactions of atom breaking is some less reactive than the radicals $R-C {\begin{smallmatrix} =O \\ O \end{smallmatrix}}$.

The stabilization energy for C_6H_5 is equal to 0,0856 $a.\ u.$, that is, its reactivity is less than for alkoxyl radicals.

Summarizing the above–said it can be concluded that the relative reactivity of alkoxyl radicals into reactions of atom breaking from any molecule will be greatly depend on stabilization energy of free radical.

On the basis of considered data it can be constructed the general range of activity for free radicals: $OH > RO > R{-}C{\lesseqgtr}^O_O > RO_2 > $ (structure with H, OCH_3, C, H, $O\cdot$) $\approx$ (structure with H, $N(CH_3)_2$, C, H, $O\cdot$).

Table 52. The stabilization energies for some radicals

№	Radical	$-E_{stab.}$ upon $O...O$ bond, a. u.	Radical	$-E_{stab.}$ upon $R{-}C$ bond, a. u.	$E_{act.,}$ kJ/ mole
1	$H_3C{-}C{\lesseqgtr}^O_O$	0,1140	$CCl_3{-}C{\lesseqgtr}^O_O$	0,9535	—
2	$C_2H_5{-}C{\lesseqgtr}^O_O$	0,1210	$CH_2Cl{-}C{\lesseqgtr}^O_O$	1,0339	—
3	$C_3H_7{-}C{\lesseqgtr}^O_O$	0,1252	$C_2H_5{-}C{\lesseqgtr}^O_O$	1,1006	—
4	$Cl{-}C{\lesseqgtr}^O_O$	0,1430	$H_3C{-}C{\lesseqgtr}^O_O$	1,1770	27,6
5	$tert.{-}C_4H_9{-}C{\lesseqgtr}^O_O$	0,1537	$C_3H_7{-}C{\lesseqgtr}^O_O$	1,1873	—
6	$CH_2Cl{-}C{\lesseqgtr}^O_O$	0,1549	$Ph{-}C{\lesseqgtr}^O_O$	1,3645	58,6
7	$Ph{-}C{\lesseqgtr}^O_O$	0,1610	—	—	—
8	$CCl_3{-}C{\lesseqgtr}^O_O$	0,1643	—	—	—

Let us consider briefly the reactions of alkoxyl radicals decomposition. For the radicals $R{-}C{\lesseqgtr}^O_{O\cdot}$ is typical the reaction of decarboxylation. From the data of Table 51 it is seen that the energy of decarboxylation activation depends not only on the strength of $R–C$ bond, but also on the value of non–valence bonds $R...O$. Unfortunately, it is not possible to construct the quantitative dependence between the experimental values of energy of decarboxylation reaction of $R–C\lessgtr^{O}_{O\bullet}$ radicals on the strength of the broken bonds with taken into account the non–valence interactions estimated under the *HMO* approximation, since in reference for the majority of reactions there are absent the experimental values of activation energy. However, accordingly to the calculations it can be assumed that the decarboxylation rate will be increased in the course of the numbers of radicals increasing (Table 52).

Let us consider the reaction of hydrogen atom detachment for radicals by *RO* type. In Table 51 there are values of energy for hydrogen atom detachment for some radicals estimated under *HMO* approximation. As we can see from the data of this table, the calculated energy is the same changed with the value of detachment enthalpy.

It is interesting the fact of methyl radical detachment from the $(CH_3)_3CO$. The calculated energy of break for the $C–C$ bond for this radical consists of 0,6784 *a. u.*, and with taken into account the interaction of *Carbon* of methyl group with oxygen atom this energy is equal to $–0,6784–0,1011 = –0,7795$ *a. u.* In a case of radical $(CH_3)_3SiO$ the value of energy for $Si–C$ bond with taken into account the interaction of *Carbon* atom with oxygen is equal to $–0,4166$ *a. u.*, that is it greatly less than for the break of bond $C–C$ in $(CH_3)_3CO$. However, for radical $(CH_3)_3SiO$ the reaction of methyl group break is not typical (Chapter 5). This is explained first of all by the competition of hydrogen atom break reaction by $(CH_3)_3SiO$ radical, since it is very reactive in these reactions upon above–said considerations. Besides, it is necessary to note, that probably into hydrogen atom break reactions by $(CH_3)_3SiO$ radical the steric factor plays the not last role, since the length of $Si–O$ bond is equal to 1,9 $\overset{0}{A}$ whereas the bond $C–O$ in $(CH_3)_3CO$ is greatly less and is equal to 1,36 $\overset{0}{A}$.

For the radicals by *RO* type it is typical the isomerization reaction with the formation of new free radicals. It is seems to us, that the more of *AO* of the nearest atom of hydrogen to the oxygen are overlapped with its *AO* and the less durable is the $C–H$ bond, the more is transition probability of this atom to the last one. Really, as we can see from the data of Table 53, the stabilization energy of radicals is antibate to the isomerization enthalpy.

It is clear, that the above–presented aspect of the reactivity estimation is more qualitative than the quantitative, since to construct of some any quantitative dependencies between the calculated magnitudes and experimental kinetic characteristics of radicals reactions in different reactions is untimely as far as this is possible only in a case of the accumulation of sufficient experimental material concerning to the structure and the reactivity of free radicals.

Table 53. The dependence of stabilization energy of radicals on isomerization enthalpy

№	Radical	Isomerization product	$-E_{stab.,}$ a. u.	$-\Delta H$ of reaction, kJ/mole
1	CH_3O	CH_2OH	0,3942	31,4
2	C_2H_5O	$CH_3–CHOH$	0,3849	39,7
3	$n–C_3H_7O$	$CH_3–CH_2–CHOH$	0,3664	—
4	$PhCH_2O$	$PhCHOH$	0,3450	46,0
5	$C_6H_5C(CH_3)_2O$	$CCH_3C_6H_5(OCH_3)$	0,8774	6,3
6	$C_6H_5C(CH_3)_2O$	$C(CH_3)_2OC_6H_5$	0,8750	6,3

INVESTIGATIONS OF KINETICS AND CHEMICAL MECHANISM OF ORGANIC PEROXIDES THERMAL DECOMPOSITION

In this Chapter the results of investigations concerning to the chemical mechanisms and kinetics of synthesized peroxides homolysis and also their quantum–chemical interpretation are considered.

5.1. REACTIVITY AND CHEMICAL MECHANISM OF TERT.– ALKYLPEROXYCYCLO-PENTENES–2 AND TERT.– ALKYLPEROXYTETRAHYDROBENZOATES THERMOLYSIS

Tert–alkylcyclopentenyl peroxides and their epoxy derivatives are compounds with bicyclic reactive center . Stabilization energy of reactive center of peroxides $-OOC(CH_3)_3$ and $-OOC(CH_3)_3$ are equal to $-1{,}0610$ *a. u.* and $-1{,}0603$ *a. u.* respectively, that corresponds to $E_{act.} \approx 140$ *kJ/mole* and is satisfactory agreed with the data of Table 54. So, these compounds upon its reactivity are little differed in homolysis reactions. The change of *tert*–butyl substitute on *tert*–amyl or *tert*–hexyl one practically does not lead to the change of the stabilization energy.

Table 54. Kinetic parameters of dialkyl peroxides thermal decomposition in solvents

№	Solvent	Peroxide	$k \cdot 10^4, s^{-1}$				
			383 K	393 K	403 K	408 K	413 K
1	n–butyl alcohol	[cyclopentene]–OOC(CH$_3$)$_3$	0,08±0,02	0,23± 0,07	—	1,10± 0,03	—
2	n–butyl alcohol	[epoxycyclohexane]–OOC(CH$_3$)$_3$	0,15± 0,04	0,46± 0,04	—	2,18± 0,04	—
3	Chloro-benzene	[cyclopentene]–OOC(CH$_3$)$_3$	—	0,22± 0,03	0,55± 0,03	—	0,62± 0,01
4	Chloro-benzene	[epoxycyclohexane]–OOC(CH$_3$)$_3$	—	0,31± 0,02	1,20± 0,01	—	3,63± 0,02
5	Dimethyl-formamide	[cyclopentene]–OOC(CH$_3$)$_3$	—	0,31± 0,02	1,19± 0,03	—	2,75± 0,06
6	Dimethyl-formamide	[epoxycyclohexane]–OOC(CH$_3$)$_3$	—	0,54± 0,05	1,89± 0,02	—	6,40± 0,02

№	E, kJ/mole	$A \cdot 10^{14}, s^{-1}$	$\Delta S^{\neq}$, e. u. (393 K)	$\Delta H^{\neq}$, kJ/mole
1	143,1 ± 0,4	2,4 ± 0,07	0,96 ± 0,25	140,1 ± 0,4
2	139,7 ± 0,4	1,83 ± 0,02	17,91 ± 0,17	136,6 ± 0,4
3	152,3 ± 0,8	43,9 ± 1,6	44,35 ± 0,12	149,2 ± 0,8
4	153,1 ± 2,5	53,9 ± 0,36	33,26 ± 0,21	141,7 ± 2,5
5	153,1 ± 1,2	67,8 ± 0,44	47,86 ± 0,20	150,0 ± 1,2
6	155,6 ± 0,4	27,1 ± 0,23	41,84 ± 1,59	152,5 ± 1,7

The primary products of $tert$–alkylperoxycyclopentenes–2 and their epoxy derivatives are free radicals [cyclopentene]–O· (*I*) and [epoxycyclohexane]–O· (*II*). The calculation shows, that for radical (*I*) more energetic conformation is the axial form than the equatorial one. The first is more advantageous than the second one on 0,02 *a. u.* (50,4 *kJ*). For free radical (*II*) there are possible four conformations – equatorial and axial «tub» and «seat» respectively. Energetically the mot advantageous is the axial «tub». The barrier of transition of this conformer into form of the equatorial «tub» is equal to 0,022 *a. u.* (55,4 *kJ*). The energies of barriers for conformer equatorial «tub» transition into equatorial and axial «seat» are equal to 0,026 *a. u.* (57,2 *kJ*) and 0,008 *a. u.* (20,2 *kJ*) respectively. These calculations show, that the transition barrier between axial conformers in 2,5 times is less than between the axial and equatorial conformers; in other words, two forms will be take part into transition reactions of ⬡–*o.* radicals under homolysis of the respective peroxides. Upon itself donor properties the substitute ⬡ in ⬡–*o.* is some stronger than ⬡ in ⬡–*o.*, since the charges on the oxygen atom are respectively equal to –0,2357 and –0,2313, and the spin densities –0,9087 and 0,9073.

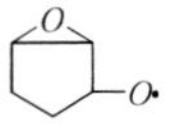

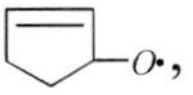

The values of stabilization energy (it is taken into account only the interactions of valence free atoms *O* and *H* into radicals ⬡ and ⬡) are little differed and respectively are equal to –0,0043 and –0,0041 *a. u.*; in other words upon themselves reactivity these radicals are equal to oxyradicals with the alkyl substitutes. $E_{stab.}$ for the last is within the ranges of –0,003 ÷ – 0,004 *a. u.*

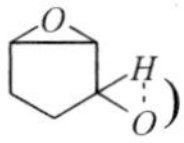

The formation of cyclopentenol–2 can be explained at the expense of hydrogen atom break from the solvent. The formation of such products as cyclopentenon and partially *tert*–butanol proceeds at the expense of the disproportionation reaction of respective radicals in cell:

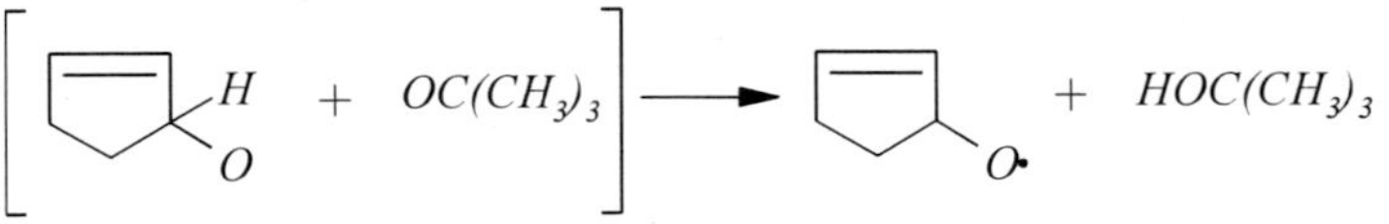

$$\left[\text{⬡}\!\!<^{H}_{O} \;+\; OC(CH_3)_3 \right] \longrightarrow \text{⬡} \!\!<_{O\cdot} \;+\; HOC(CH_3)_3$$

$$(21)$$

This assumption is confirmed by relatively little outcome of radicals from the cell ($e = 0,3$) under thermolysis of the corresponding peroxide.

Forcible argument of the cell–like formation of cyclopentenon is its changed accumulation in the presence of inhibitor. Formation of this product could be explained by the reaction

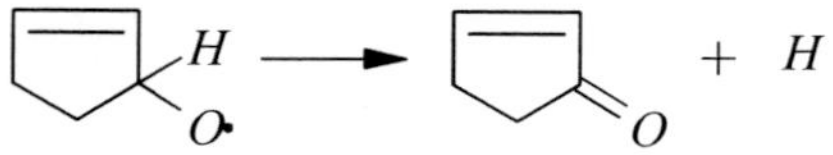

$$(22)$$

However, hydrogen is absent under the cyclopentenyl peroxides decomposition.

Some less quantity of the forming epoxycyclopentanone under the epoxycyclopentane peroxides decomposition confirms the decrease of the percentage of disproportionation reaction of respecting radicals and increase of the $\qquad$ and $OC(CH_3)_3$ radicals outcome from the cell. Thereof was expecting, since the conformations of free radicals $\qquad$ and $\qquad$ are absolutely differed. The axial «tub» is typical for the first radical, and for the second radical is typical the axial form. So, the outcomes of these radicals from a cell should be differed. Really, the outcome of radicals from a cell for peroxide $\qquad$ $OOC(CH_3)_3$ within the temperature interval $413 - 433\ K$ is equal to $0,3 - 0,4$ whereas for the $\qquad$ $OOC(CH_3)_3$ this value consists of $0,4 - 0,6$ *(see Table 55)*.

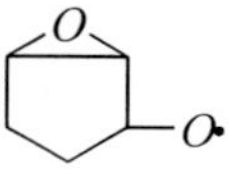

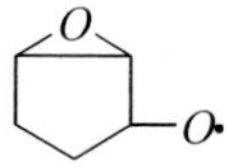

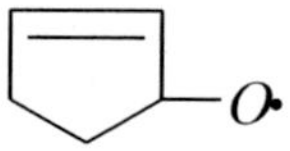

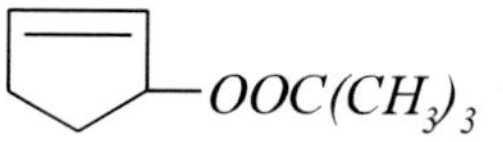

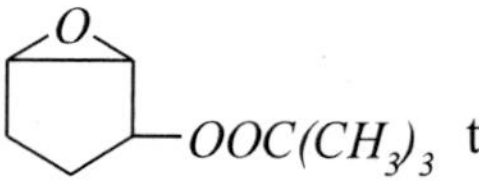

Based on the above presented considerations the thermolysis of *tert–alkylperoxycyclopentenes–2* and their epoxy derivatives can be represented by the following scheme:

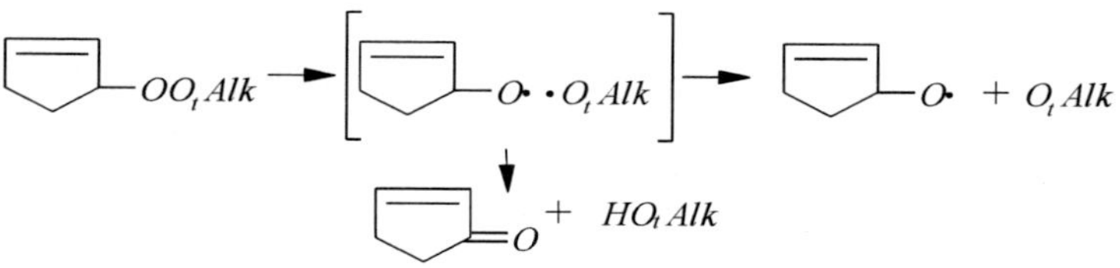

$$(23)$$

**Table 55. Kinetic parameters of *tert*–alkylperoxycyclopentenes–2
and their epoxy derivatives thermolysis**

№	Compound	T, K	$k_{decomp.} \cdot 10^4, s^{-1}$	$E_{act.}$, kJ/mole	e, outcome of radicals from cell
1	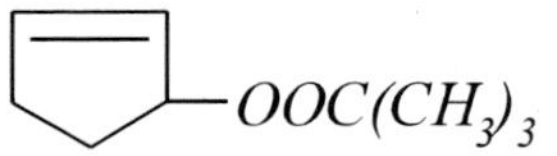 —$OOC(CH_3)_3$	413	1,4	147,3 ± 2,1	0,30
		423	3,3		0,32
		433	8,5		0,40
2	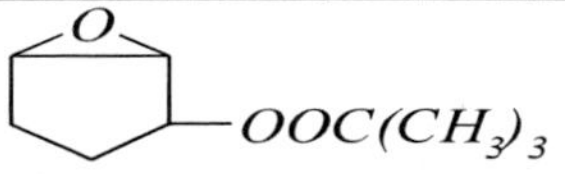 O—$OOC(CH_3)_3$	413	1,8	145,2 ± 3,3	0,46
		423	4,3		0,47
		433	12,2		0,57
3	 —$OOC(CH_3)_2C_3H_7$	413	1,4	141,0 ± 4,2	0,32
		423	4,0		0,38
		433	8,7		0,43
4	 O—$OOC(CH_3)_2C_3H_7$	413	2,2	153,1 ± 2,5	0,60
		423	5,6		0,66
		433	14,4		0,68

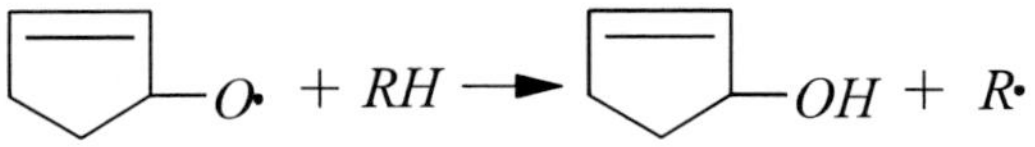

$$\text{—}O\bullet + RH \longrightarrow \text{—}OH + R\bullet \tag{24}$$

$$\bullet O_t Alk \longrightarrow product + R_1 \tag{25}$$

$$O_t Alk + RH \longrightarrow HO_t Alk \tag{26}$$

$$R_1^{\bullet} + RH \rightarrow R_1 H + R^{\bullet} \tag{27}$$

Peresters $\text{—}\overset{O}{C}\text{—}O\text{–}O\text{–}C(CH_3)_3$ and $O\text{—}\overset{O}{C}\text{—}O\text{–}O\text{–}C(CH_3)_3$ are

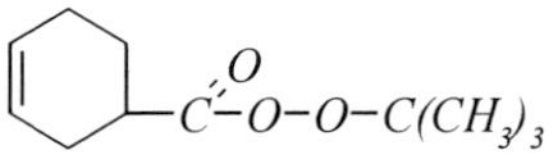

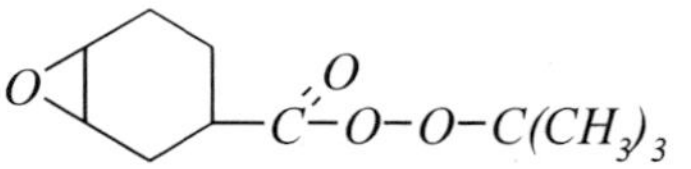

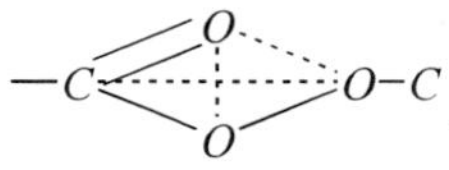

compounds with polycyclic reactive center $-C\overset{O}{\underset{O}{\cdots}}O-C$. Stabilization energies of these compounds are equal to $-0,8870$ *a. u.* and $-0,8874$ *a. u.*, that corresponds to the activation energy of homolysis 125 *kJ/mole* and is satisfactory agreed with the experimental values of activation energy. For the primary oxyradicals 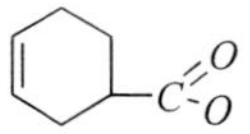 (*I*) and 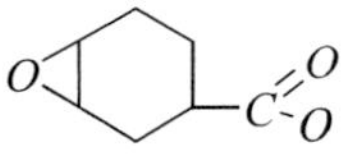 (*II*) the charge on atoms of oxygen is delocalized and equal to $-0,3969$ and $-0,3987$ respectively. This is some higher than for *PO* with alkyl substitutes. The spin densities for (*I*) and (*II*) are equal to 0,3615 and 0,3608 respectively. These characteristics are near to the characteristics for $(CH_3)_3CC$ radical. The stabilization energies for radicals (*I*) and (*II*) upon interactions of valence free atoms of oxygen C are equal to $-0,1534$ *a. u.* and $-0,1531$ *a. u.* It can be considered that their reactivity in reactions of hydrogen atom break or in reactions of addition upon double bonds is practically the same. The stabilization energy of decarboxylation (taking into account the strength of *C–C* bond) of 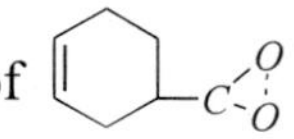 radical ($-1,1868$ *a. u.*) is in a field of stabilization energy of alkoxyl radicals ($-1,10 \div 1,19$ *a. u.*).

Peroxides $OCH_2OO-C(CH_3)_3$ and $OCH_2OO-C(CH_3)_3$ are the compounds with polycyclic reactive center; stabilization energy for them are practically the same and are equal to $-0,8864$ *a. u.* and $-0,9086$ *a. u.* with taken into account the valence free atoms of center; this corresponds to the activation energy of homolysis in chlorobenzene: $\sim$ 123 *kJ/mole* and $\sim$ 132 *kJ/mole*. The primary products of homolysis, namely radicals $C-OCH_2O\bullet$ and $C-OCH_2O\bullet$ are disposed to the hydrogen atom detachment with the formation of compound *RC(O)OC(O)H*.

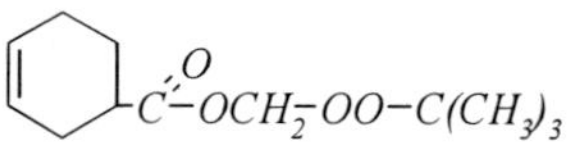

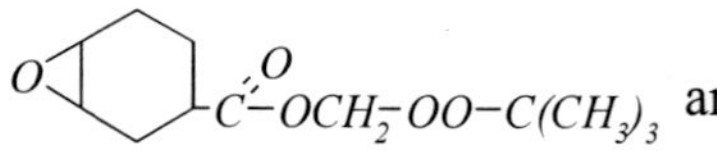

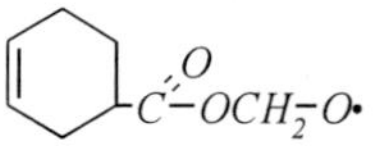

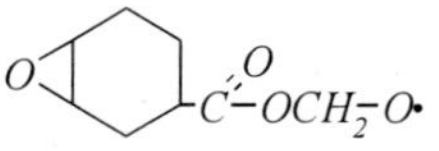

Radical 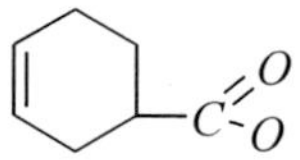 is easy decarboxylated in dichlorobenzene. In tetradecane and nonyl alcohol the 50 % of tetrahydrobenzoic acid is formed, that is this radical detaches the hydrogen atom mainly from the solvent.

The formation of cyclohexadiene could consider also by cell–like product, since it is formed also in the presence of α–naphtylamine. Possibly, some part of it is formed also as a result of induced decomposition of investigated peroxide or hydrogen atom detachment from radical [structure]. Most likely that the last variant is realized, since it explains the formation of hydrogen at the thermolysis.

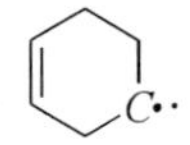

Taking into account the above–said, the decomposition of *tert*–alkylperoxytetrahydrobenzoates and their epoxy derivatives can be represented by the following scheme:

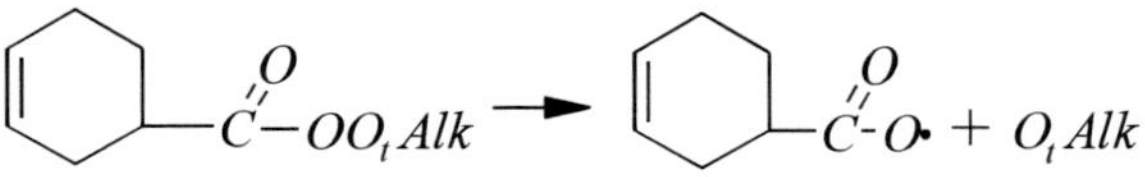

$$[\text{structure}]-\overset{O}{\underset{\|}{C}}-OO_tAlk \longrightarrow [\text{structure}]-\overset{O}{\underset{\|}{C}}-O\cdot + O_tAlk \tag{28}$$

$$[\text{structure}]-\overset{O}{\underset{\|}{C}}-O\cdot \longrightarrow [\text{structure}]C\cdot + CO_2 \tag{29}$$

$$[\text{structure}]C\cdot + RH \longrightarrow [\text{structure}] + R \tag{30}$$

$$[\text{structure}]-\overset{O}{\underset{\|}{C}}-O\cdot + RH \longrightarrow [\text{structure}]-\overset{O}{\underset{\|}{C}}-OH + R \tag{31}$$

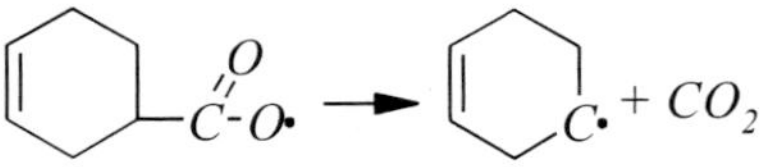

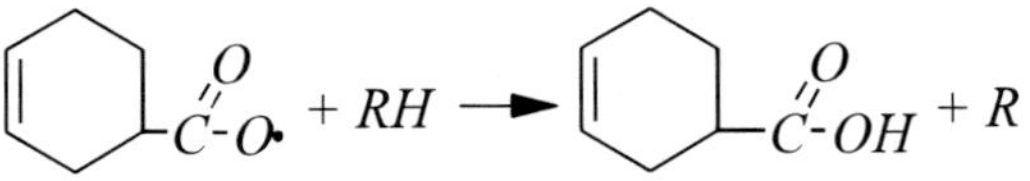

$$O_tAlk \longrightarrow R^\bullet + product \tag{32}$$

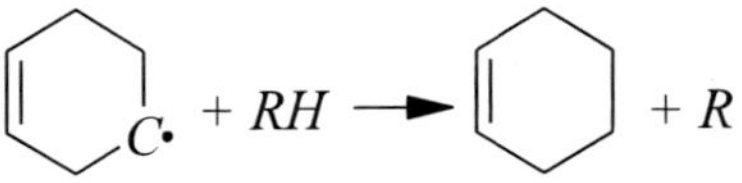

$$O_tAlk + RH \longrightarrow R_1H + R \tag{33}$$

$$R_1^\bullet + RH \longrightarrow R_1H + R^\bullet \tag{34}$$

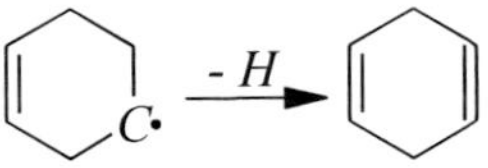

$$[\text{structure}]C\cdot \overset{-H}{\longrightarrow} [\text{structure}] \tag{35}$$

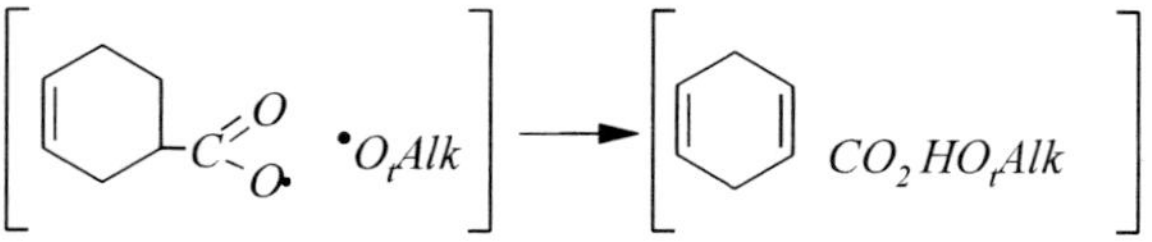

$$\left[\underset{C\overset{O}{\underset{O\cdot}{\diagup}}}{\bigodot} \;\; {}^{\bullet}O_tAlk \right] \longrightarrow \left[\bigodot \;\; CO_2\ HO_tAlk \right] \tag{36}$$

In the beginning of this Chapter it was already accentuated that the sufficient low value of the activation energy of compound

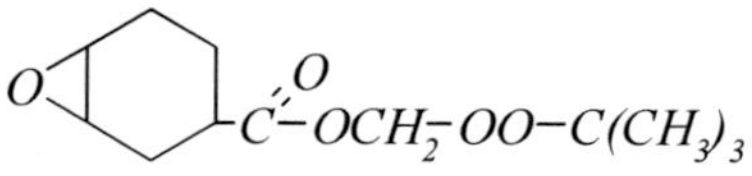

$$O\!\!\!\triangleleft\!\!\bigodot\!-C\overset{O}{\underset{}{\diagup}}-OCH_2^--OO-C(CH_3)_3$$ decomposition (this compound most likely

it is necessary to classify as the substituted dialkyl peroxides) is explained by the specificity in structure of its reactive center like to the center of peresters or nitrogen–containing peroxides.

A much of hydrogen and formyl–3,4–epoxyhexahydrobenzoate are formed under thermal decomposition of the above mentioned peroxide. These two products are result of the hydrogen atom detachment from the radical

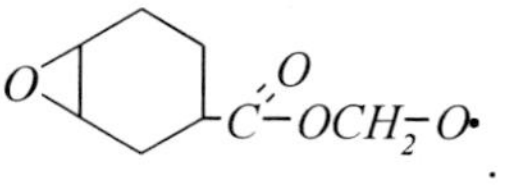

$$O\!\!\!\triangleleft\!\!\bigodot\!-C\overset{O}{\underset{}{\diagup}}-OCH_2^--O\cdot$$

.

5.2. THERMAL DECOMPOSITION OF TERT–ALKYLPEROXYMETHYLAMINES

It was studied the thermal decomposition of the following compounds:

$$(C_2H_5)_2NCH_2OOC(CH_3)_3 \tag{I}$$

$$C_5H_{10}NCH_2OOC(CH_3)_3 \tag{II}$$

$$(CH_3)_2NCH_2OOC(CH_3)_3 \tag{III}$$

$$(iso–C_3H_7)_2NCH_2OOC(CH_3)_2C_2H_5 \tag{IV}$$

$$(CH_3)_2NCH_2OOC(CH_3)_2C_2H_5 \tag{V}$$

Kinetic parameters of the nitrogen–containing peroxides (*I*)–(*V*) decomposition have been studied in references [91–93]. In all cases at little starting concentrations of these peroxides the kinetics of their thermolysis is described by the first order equation. However, at increasing of the starting concentrations of peroxides it is observed the essential deviations from this equation. On *Figure* 42 there are kinetic curvers of the peroxide (*I*) decomposition in chlorobenzene in semilogarithmic representation. Now at concentration 0,03 *mole/l* it is observed the sharp deviation from the first order equation. Probably, the percentage of the induced decomposition takes place here. The same dependence of decomposition rate constant on the concentration for peroxide (*I*) is observed also at its thermal decomposition into *n*–amyl alcohol. On *Figure* 43 there is dependence of decomposition rate constant for peroxide (*IV*) on the starting concentrations at different temperatures into chlorobenzene and *n*–amyl alcohol. For presented peroxide the equation of the first order is good realized via decomposition reaction proceeding, but as we can see from this figure it is observed the dependence of decomposition rate constant on starting concentration of peroxide. The values of decomposition rates constants obtained via the extrapolation of k dependence on starting concentration to its zero value *(see Figure* 43*)* are in good agreement with the decomposition rates constants for (*IV*) peroxide at the same temperature and in the same solvents in the presence of inhibitor α–naphthylamine.

In references [94–99] it was studied the kinetics of decomposition both for mono– and bifunctional nitrogen–containing peroxides by inhibition and chemiluminescence methods. In all cases it was observed too much low values of activation energies of pre–exponential factors, which are not typical for the monomolecular decomposition of peroxides. For example, the inhibition method for nitrogen–containing peroxides gives the value $E_{act.}$ equal to 71–104 *kJ/mole* depending on the solvent nature. Solvents able to the specific solvation decrease of this characteristic. Low are also the pre–exponents values ($10^6 - 10^{10}\ s^{-1}$). The understated kinetic parameters can be explained by the induced reactions forming by the radicals from inhibitor and by the molecules of peroxides.

The chemiluminescence method also gives the understated kinetic parameters of nitrogen–containing peroxides decomposition. Probably, first of all it is related with some secondary reactions of forming radicals leading to the additional chemiluminescence.

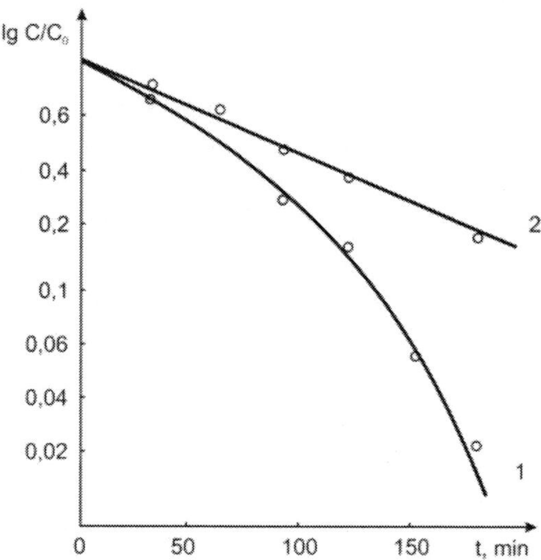

Figure 42. The kinetic curvers of diethylaminemethyl–*tert*–butylperoxide decomposition in chlorobenzene at starting concentrations: (1) – 0,03 *mole/l*; (2) – 0,005 *mole/l*; C and C_0 are current and starting concentration of peroxide respectively; $T = 399\ K$. Semilogarithmic representation.

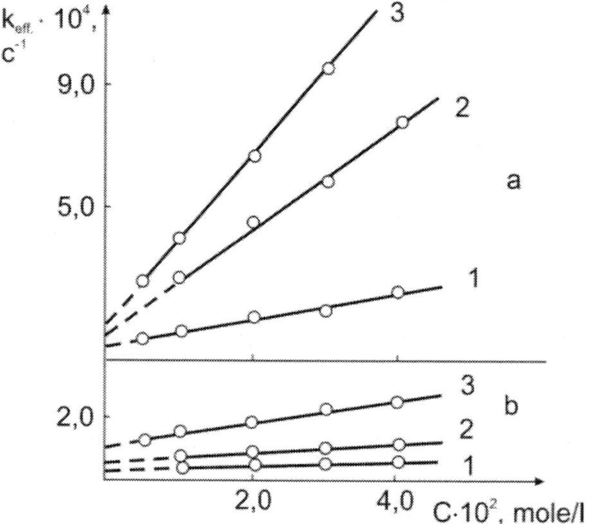

Figure 43. Dependence of the diisopropylaminemethyl–*tert*–amylperoxide decomposition rate constant on starting concentration at different temperatures. *(a)* – *n*–amyl alcohol: (1) – 393 K; (2) – 403 K; (3) – 408 K. *(b)* – chlorobenzene: (1) – 381 K; (2) – 389 K; (3) – 398 K.

That is why it was made an attempt to study the kinetics of compounds (I) – (V) thermolysis in medium of styrene in the presence of α–naphthylamine. Styrene is sufficiently good acceptor of radicals. The α–naphthylamine prevented the polymerization process of styrene. The thermal decomposition in styrene in the presence of inhibitor practically excludes the all types of the induced decomposition of peroxides and the constant of process should be maximally neared to the constant of the monomolecular decomposition.

Judging upon the final products of the thermal decomposition reactions of nitrogen–containing peroxides in chlorobenzene and neglecting by the disproportionation reaction $R_1R_1NCH_2O^{\cdot} + {}^{\cdot}O_tAlk \rightarrow R_1R_1NCHO + HO_tAlk$ it can be written the following scheme:

$$R_1R_1NCH_2OO_tAlk \xrightarrow{\ k_1\ } R_1R_1NCH_2O^{\cdot} + {}^{\cdot}O_tAlk \tag{37}$$

$$R_1R_1NCH_2O \xrightarrow{\ k_2\ } H + R_1R_1NC\underset{H}{\overset{O}{\diagup}} \tag{38}$$

$$R_1R_1NCH_2OO_tAlk + O_tAlk \xrightarrow{\ k_3\ } \tag{39}$$

$$\longrightarrow R_1R_1NC\underset{H}{\overset{O}{\diagup}} + HO_tAlk + O_tAlk \tag{40}$$

$$H + {}^{\cdot}O_tAlk \xrightarrow{\ k_1\ } HO_tAlk \tag{41}$$

It is necessary to note, that the proposed scheme takes place under thermal decomposition of nitrogen–containing peroxides (I) – (V) decomposition in relatively inert solvents (chlorobenzene and others). Under the decomposition of presented peroxides in n–amyl alcohol this scheme is very complicated.

5.3. ABOUT CHEMICAL MECHANISM OF NITROGEN–CONTAINING PEROXIDES THERMAL DECOMPOSITION

Investigation of nitrogen–containing thermal decomposition kinetics leads to conclusion that the chemical mechanism of these compounds decomposition it is enough complicated and have the radical–chain character. Nitrogen–containing peroxides by general formula $R_1R_1N–CH_2OOR$ are the compounds with polycyclic reactive center.

Stabilization energy of the center's atoms has a little dependence on the R nature (CH_3, C_2H_5, C_3H_7) and is equal to $-0,9203 \pm -0,9210$ *a. u.* for these compounds. Quite naturally, that the homolysis activation energies of these compounds are little differed between themselves (130–134 *kJ/mole*).

Our considerations about the chemical mechanism are based on kinetic data of the chromatographic analysis of the final products of the peroxides decomposition and also on quantum–chemical calculations both of molecules and forming free radicals.

It can be supposed, that based on kinetic data (sufficiently low values $E \sim$ 130–134 *kJ/mole*) the nitrogen–containing peroxides are decomposed synchronously, in other words with the simultaneous break of $-O–O-$ and $-C–H-$ bonds of methylene bridge. However, such supposition is not wealthy, since the specially carried out quantum–chemical calculation shows, that the synchronous break is less energetically advantageous than the break upon $-O–O-$ bond only.

From the other hand, purely experimental fact (a little quantity of the forming hydrogen) also contradicts to the assumption about the synchronous break of the bonds in these peroxides.

It's follows from this, that probably the first act into thermolysis reactions for nitrogen–containing peroxides is the break of $-O–O-$ bonds with the formation of two free radicals. In general, forming radical $R_1R_1NCH_2O$ accordingly to our imaginations is inactive in reactions of the hydrogen break and is more disposed to the reaction with the hydrogen detachment with the formation of molecular product dialkylformamide.

Let us consider this question in more detail. The radical $(CH_3)_2CHCH_2O$ will be the comparison radical. So, radicals detaching the hydrogen form the respective aldehydes:

$$(CH_3)_2 N\overset{\overset{H}{+}}{\underset{H}{C}}-O\cdot \longrightarrow (CH_3)_2 NC\overset{O}{\underset{H}{\diagdown}} \tag{42}$$

$$(CH_3)_2 CH-\overset{\overset{H}{+}}{\underset{H}{C}}-O\cdot \longrightarrow (CH_3)_2 CH-C\overset{O}{\underset{H}{\diagdown}} \tag{43}$$

The energy of hydrogen atom detachment for reaction (42) is equal to:

$$Q_1 = -55{,}4575 - (-54{,}6262 + (-0{,}6387)) = -0{,}1926 \; a.\,u. = -505{,}4 \; kJ$$

and for reaction (43) is equal to:

$$Q_2 = -51{,}9417 - (-51{,}0983 + (-0{,}6387)) = -0{,}2047 \; a.\,u. = -537{,}2 \; kJ$$

It is follows from the calculations, that the hydrogen detachment from $(CH_3)_2NCH_2O$ is more advantageous, than in a case of the same reaction for $(CH_3)_2CHCH_2O$.

Investigation of kinetics of the styrene polymerization process initiated by nitrogen–containing peroxides permits to consider about the low reactivity of $R_1R_1NCH_2O$ in addition reactions also. This free radical is weakly added to the molecule of styrene.

Specially carried out experiment confirms the inability of $R_1R_1NCH_2O$ to detach the hydrogen from the molecules of solvent. The essence of an experiment is in the following: the thermal decomposition of $(CH_3)_2NCH_2OOC(CH_3)_3$ was carried out in the presence of dimethylamine exceed in chlorobenzene and ethylbenzene. Let us assume, that in accordance with the reaction $(CH_3)_2NCH_2O + RH \rightarrow (CH_3)_2NCH_2OH + R$ the dimethylaminomethanol is formed. This compound is an intermediate product of the *Mannikh's* reaction [100] and is unstable. However, in the presence of dimethylamine the tetramethylmethylendiamine should be formed accordingly to reaction:

$$(CH_3)_2NCH_2OH + HN(CH_3)_2 \rightarrow (CH_3)_2NCH_2N(CH_3)_2 + H_2O \tag{44}$$

The reaction (44) easy proceeds and is one among stages of the reaction of formaldehyde condensation with the secondary amines in accordance with scheme:

$$2(CH_3)_2NH + HCHO \rightarrow (CH_3)_2NCH_2N(CH_3)_2 + H_2O \qquad (45)$$

However, this product under the peroxide thermolysis in the presence of dimethylamine is not formed. This experiment sufficiently earnestly confirms our supposition about fact, that the radical $R_1R_1NCH_2O$ can not break the hydrogen from the molecule of a solvent or induce the process of thermal decomposition of investigated peroxides. The main products of the nitrogen–containing peroxides decomposition are the dialkylformamide and the products of the *tert*–alkoxyl radical transitions reactions.

Dialkylformamides are the products of the radical $R_1R_1NCH_2O^\cdot$ transition reactions and the induced decomposition of the nitrogen–containing peroxides:

$$R_1R_1NCH_2OO_tAlk + (R, \overset{\bullet}{O}_tAlk) \longrightarrow R_1R_1NCHOO_tAlk$$

$$R_1R_1NC\overset{\nearrow O}{\underset{\searrow H}{}} \quad + \quad O_tAlk \qquad (46)$$

The formation of 2–alkylpropene–1 oxide the best of all is explained by the intermolecular reaction $S_{N2} - \beta$–cleavage of $-O-O-$ bond of dialkylaminomethyl–*tert*–alkylperoxide radical, which is formed as a result of the hydrogen atom break from the peroxide R or O_tAlk by radicals:

$$R_1R_1NCH_2OOC\overset{\overset{\bullet}{C}H_2}{\underset{}{}}-(R)_2 \longrightarrow R_1R_1NCH_2O + H_2-C\overset{(R_1)_2-C}{\underset{}{}}O \qquad (47)$$

The formation of ethane is explained by the methyl radicals recombination reaction.

A plenty of forming *tert*–alkyl alcohols under nitrogen–containing peroxides decomposition in chlorobenzene keeps to take into consideration the reaction:

$$R_1R_1NCH_2OO_tAlk + O_tAlk \rightarrow R_1R_1NCHOO_tAlk + HO_tAlk \qquad (48)$$

$$R_1R_1NCHOO_tAlk \longrightarrow R_1R_1NC{\overset{\displaystyle O}{\underset{H}{\diagup\diagdown}}} + O_tAlk \qquad (49)$$

and to take into consideration the reaction $O_tAlk + solvent \rightarrow HO_tAlk$.

Based on the theoretical and experimental imaginations about thermolysis of the nitrogen–containing peroxides it can be proposed the following scheme of their decomposition:

$$R_1R_1NCH_2OO_tAlk \rightarrow R_1R_1NCH_2O^{\cdot} + O_tAlk \qquad (50)$$

$$R_1R_1NCH_2O \longrightarrow R_1R_1NC{\overset{\displaystyle O}{\underset{H}{\diagup\diagdown}}} + H \qquad (51)$$

$$R_1R_1NCH_2O^{\cdot} + O_tAlk \rightarrow R_1R_1NCHO + HO_tAlk \qquad (52)$$

$$R_1R_1NCH_2OO_tAlk + O_tAlk \rightarrow R_1R_1NCHO + HO_tAlk + O_tAlk \qquad (53)$$

$$O_tAlk \rightarrow inactive\ product + R^{\cdot} \qquad (54)$$

$$R_1R_1NCH_2OO_tAlk \xrightarrow{\ R_1H\ } R_1R_1NCHO + (R_1 + H_2) + O_tAlk \qquad (55)$$

$$R^{\cdot} + R^{\cdot} \rightarrow RR \qquad (56)$$

$$O_tAlk + O_tAlk \rightarrow inactive\ product \qquad (57)$$

$$H + O_tAlk \rightarrow HO_tAlk \qquad (58)$$

Depending on the nature of solvents in which the peroxide thermolysis proceeds and also on quantity of the final products, some of the above presented stages can be neglected. However, the neglecting by other stages needs the experimental evidence. For example, under $R_1R_1NCH_2OO_tAlk$

decomposition in chlorobenzene and ethylbenzene it is formed not enough hydrogen. In such a case the stages (51) and (55) can be rejected, proved at this that tert–butanol is not formed accordingly to reaction (58).

It is seems to us, that the disproportionation reaction (52) is not main under the respective products formation since the radicals outcome from a cell under peroxides thermolysis is enough great (~1) and, from the other hand, if this fact would had an essential place, then hardly the nitrogen–containing peroxides would effectively initiate the styrene polymerization process.

5.4. KINETICS AND CHEMICAL MECHANISM OF ORGANOSILICON PEROXIDES THERMAL DECOMPOSITION

Some aspects of the relationship between the reactivity of organosilicon peroxides and their structure in thermolysis reactions are considered in this *Subsection.*

Kinetics and chemical mechanism of the organosilicon peroxides thermal decomposition were in detail studied in works [100–105].

It has been studied the following compounds:

Tert–butyltrimethylsilylperoxide *(CH₃)₃COOSi(CH₃)₃*   (*I*);

Hexamethyldisilylperoxide *(CH₃)₃SiOOSi(CH₃)₃*   (*II*);

Tert–butyldimethyl(methoxy)–silylperoxide *(CH₃)₃*
        COOSi(CH₃)₂CH₃O   (*III*).

Kinetics of the compounds (*I*) – (*III*) thermal decomposition in a range of the concentrations $8 \cdot 10^{-3}$ – $8 \cdot 10^{-2}$ *mole/l* is described by the first order equation and does not depend on the concentration of peroxides *(Figures 44–45)*.

Organosilicon peroxides are enough thermally stable compounds. An introduction of the one atom of *Silicon* into *tert*–butyl peroxide already considerably increases its thermal stability. Hexamethyldisilylperoxide is some more stable compound.

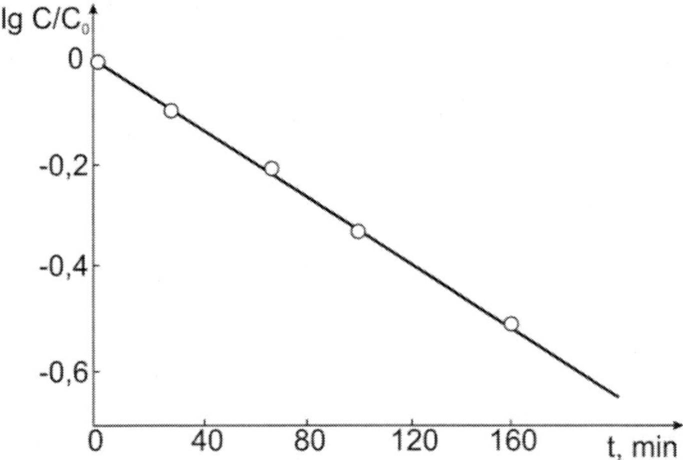

Figure 44. Dependence of $lg\dfrac{c}{c_0}$ on time under thermolysis of *(CH₃)₃COOSi(CH₃)₃*. *T* = 463 *K*. Solvent – chlorobenzene.

The main products of the peroxides (*I*) – (*III*) thermal decomposition are represented in Table 56.

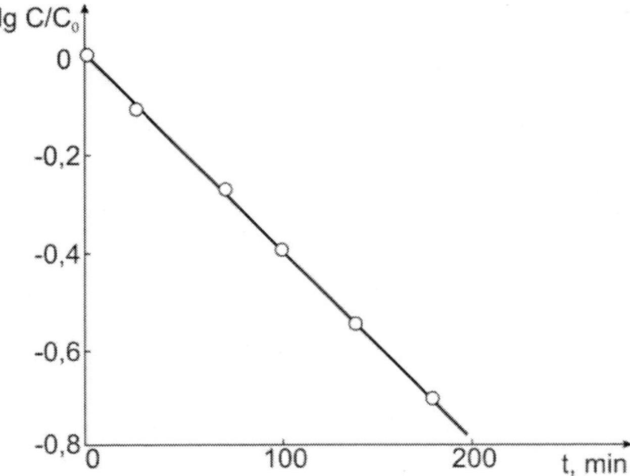

Figure 45. Dependence of $lg\dfrac{c}{c_0}$ on time under thermolysis of *(CH₃)₃SiOOSi(CH₃)₃*. *T* = 463 *K*. Solvent – chlorobenzene.

Table 56. The products of the organosilicon peroxides decomposition
(mole/mole of the decomposed peroxide)

Peroxide	The products of decomposition	Dimethylp-htalate	Tetra-decane	Chloro-benzene
(*I*)	Acetone	0,60	0,29	0,50
	Tert–butyl alcohol	0,40	0,64	0,26
	Hexamethyldisiloxane	0,45	0,42	0,42
	Silanol	0,09	0,18	0,16
	Methane	0,58	0,28	0,48
(*II*)	Hexamethyldisiloxane	—	0,47	0,46
	Silanol	—	0,08	0,04
(*III*)	Acetone	1,28	1,41	1,40
	Tert–butyl alcohol	0,72	1,38	0,51
	Methane	1,17	0,42	1,12

The transformation product of the free radical $(CH_3)_3SiO$ is the silanol $(CH_3)_3SiOH$, which is easy dimerizated into hexamethyldisiloxane. Starting from the products of compounds (*I*) – (*III*) decomposition it is proposed the following scheme of the organosilicon peroxides decomposition:

$$(CH_3)_3SiOOSi(CH_3)_3 \rightarrow 2(CH_3)_3SiO^\cdot \qquad\qquad (59)$$

$$(CH_3)_3SiO + RH \rightarrow (CH_3)_3SiOH \qquad\qquad (60)$$

$$2(CH_3)_3SiOH \rightarrow (CH_3)_3SiOSi(CH_3)_3 + H_2O \qquad\qquad (61)$$

Taking into account the products of a decomposition, the radical $(CH_3)_3SiO^\cdot$ is very reactive into reactions of the hydrogen atom break from the solvents. It were determined the rate constants of the peroxide (*I*) thermolysis by inhibition method. The decomposition rates constants are equal to $2,4 \cdot 10^{-4}$ s^{-1}, $5,9 \cdot 10^{-4}$ s^{-1} and $14,5 \cdot 10^{-4}$ s^{-1} at temperatures 463, 478 and 493 K respectively. Based on the peroxide consumption without the inhibitor the rates constants at the same temperatures are equal to $1,3 \cdot 10^{-4}$ s^{-1}, $4,3 \cdot 10^{-4}$ s^{-1} and $14,5 \cdot 10^{-4}$ s^{-1} respectively. An operating concentration of α–naphtylamine was equal to $5 \cdot 10^{-3}$ $mole/l$.

The rates constants of the peroxide thermal decomposition in the presence of the inhibitor exceed are practically as same as at its lower concentrations.

For example, for peroxide (*I*) at the concentrations of inhibitor $3 \cdot 10^{-2}$ *mole/l* and $1 \cdot 10^{-1}$ *mole/l* the rates constants at 473 *K* are equal to $2{,}85 \cdot 10^{-4}$ s^{-1}, $2{,}64 \cdot 10^{-4}$ s^{-1} and $2{,}60 \cdot 10^{-4}$ s^{-1}. This means, that under presented conditions the inhibitor is consumed only as the acceptor of free radicals [106].

High thermal stability of the organosilicon peroxides is explained by a great energy of the stabilization system, which upon its value is considerably higher, than for the usual dialkylperoxides. For organosilicon peroxides the stabilization energy is increased at the expense of too much non–valence interactions of oxygen atoms with the unneighbouring atoms of *Silicon*.

5.5. ABOUT SOME ASPECTS OF THE SOLVENT NATURE INFLUENCE ON DECOMPOSITION KINETICS OF ORGANIC PEROXIDES

At the analysis of kinetics of *tert*–alkylperoxycyclopentenes and *tert*–alkylperoxytetrahydrobenzoates thermolysis in different solvents it is necessary to take into account the universal and specific influences *(see* Table 54*)*. Let us consider this question in some detail. Unfortunately, the majority of phenomena in liquid kinetics, the behaviour of particles into solution can not be described from the invisible positions [107–111].

An analysis of obtained data *(see* Table 54*)* confirms the fact that the nature of solvent has an influence on kinetic parameters of the investigated peroxy compounds decomposition. Thus, the comparison of the activation energy magnitudes for the reactions of peroxy compounds decomposition in different solvents show, that in a case of *n*–butanol it is legibly retraced the tendency towards decreasing of this value. In a case of the peroxides decomposition in solvents: chlorobenzene ($\varepsilon = 5{,}612$), nitrobenzene (($\varepsilon = 34{,}62$), dimethylformamide ($\varepsilon = 37{,}65$) and styrene ($\varepsilon = 2{,}4252$) differing by the dielectric permeabilities, it is observed some insignificant differences into activation energy values. Evidently, as same as in a case of the peroxyacetals a dielectric permeability of medium hasn't the essential influence on the activation energy of the investigated peroxides thermolysis reaction and the differences into the values of activation energies can be refereed to the experiment error.

It has been investigated the dependence between the peroxide decomposition rate on example of *tert*–butyltetrahydroperbenzoate and values E_T proposed by Dimrott and Reinhardt [108]. In this connection it was shown that for chlorobenzene, dimethylformamide and nitrobenzene it was observed the linear dependence of the decomposition rate constant on E_T *(see Figure* 46) and the value for butanol is fallen out from this dependence. From our point of view the last is caused by the specific solvation.

Determined in such way facts can be considered as an argument in favour of that the medium polarity influence on kinetics of the peroxides decomposition in butanol is graded by the formation of hydrogen bonds between the peroxy compound and alcohol. The evidences of such relationship existence are represented below.

Some other regularity is observed at the consideration of the solvent nature influence on the activation entropy values of the peroxides thermolysis. In a case of the asymmetrical dialkyl peroxides, *namely tert–*butylperoxycyclopentene–2 and *tert–*butyl–2,3–epoxycyclopentenylperoxide the lowest activation entropy was in butanol. In dimethylformamide (ε = 37,65) and in chlorobenzene (ε = 5,612) which are visible differed by the dielectric permeability, the activation entropies are differed insignificantly. Evidently, in presented case the dielectric permeability of medium hasn't the essential influence on the activation entropy change, whereas in a case of the alcohol these changes in comparison with the dimethylformamide and chlorobenzene are enough great. First of all this is caused by the specific solvation peroxide–alcohol. Besides, evidently, the entropies of the transition state and of the starting system are little differed. The last is caused by fact that the freedom of the transition state in consequence of the solvation is closely approximated from the freedom of the starting state.

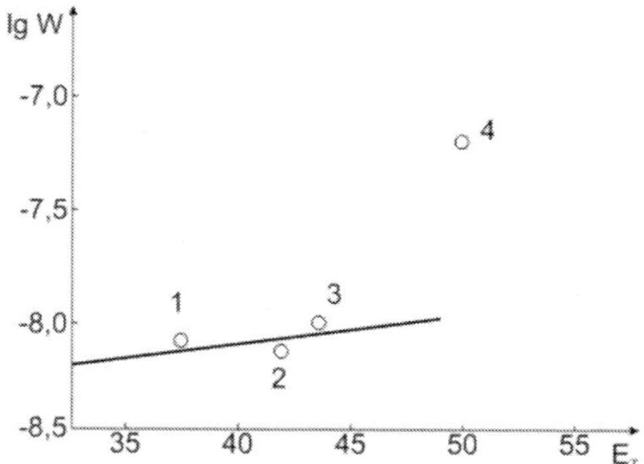

Figure 46. The influence of solvent on decomposition rate of *tert–*butyltetrahydroperbenzoate at 298 *K. 1 –* chlorobenzene; *2 –* nitrobenzene; *3 –* dimethylformamide; *4 – n–*butyl alcohol.

Table 57. Kinetic parameters of *tert*–butyl–3,4–epoxyhexahydroperbenzoate and *tert*–butylperoxymethyl–3,4–epoxyhexahydrobenzoate decomposition in mixture of different solvents

№	Solvent, % vol.			ε, 368 K	$k\cdot10^4, s^{-1}$					E, kJ/mole
	buta-nol	chloro-benzene	sty-rene		358	368	378	388	400	
1	100	—	—	—	1,39±0,02	4,12±0,89	8,27±0,11	—	—	111,3±1,7
2	95	5	—	9,86	—	3,85±0,03	6,83±0,09	19,67±0,11	—	105,8±0,8
3	70	30	—	8,74	—	2,13±0,02	4,54±0,03	14,48±0,16	—	113,4±0,8
4	50	50	—	7,75	—	1,18±0,03	3,02±0,01	9,62±0,37	—	123,4±0,5
5	20	80	—	5,97	—	0,69±0,04	—	2,92±0,07	18,2±0,15	118,4±1,5
6	5	95	—	4,93	—	0,72±0,08	2,16±0,02	6,49±0,16	—	127,6±2,6
7	—	100	—	—	—	0,64±0,00	1,82±0,01	5,59±0,07	—	127,2±0,8
8	20	60	20	—	—	0,71±0,02	—	3,00±0,03	21,38±0,40	128,0±1,8

№	Solvent, % vol.			ε, 368 K	$k\cdot10^4, s^{-1}$					E, kJ/mole
	buta-nol	chloro-benzene	sty-rene		378	388	398	403	413	
1	100	—	—	—	0,84±0,02	2,65±0,02	2,65±0,07	—	—	117,6±1,7
2	95	5	—	9,86	0,92±0,04	2,59±0,05	5,82±0,08	—	—	110,4±0,8
3	70	30	—	8,74	0,66±0,03	2,23±0,04	5,12±0,12	—	—	114,6±2,1
4	50	50	—	7,75	—	—	—	—	—	—
5	20	80	—	5,97	0,42±0,02	1,40±0,02	3,79±0,05	—	—	113,4±0,8
6	5	95	—	4,93	0,34±0,03	1,15±0,09	2,63±0,03	—16	—	117,2±2,1
7	—	100	—	—	—	0,83±0,02	1,74±0,04	4,24±0,05	—	132,2±4,2
8	20	60	20	—	0,61±0,03	1,79±0,09	3,69±0,08	—	—	111,7±0,5

Analysing the solvents nature influence on the activation entropy in a case of peresters it is necessary to note, that the essential influence has only an alcohol in view of above–said considerations. It is difficult to say something more determinantal about the influence of other solvents since the activation entropies in these media are differed insignificantly. Besides, the experimental determination of the pre–exponent is conjugated with the enough great errors. However, for *tert*–butylperoxy and *tert*–butylperoxymethyl esters of tetrahydrobenzoic acid and their epoxy derivatives it is retraced the following tendency into activation entropy change: $\Delta S^{\neq}$ (butanol) $< \Delta S^{\neq}$ (styrene) $< \Delta S^{\neq}$ (dimethylformamide) $< \Delta S^{\neq}$ (nitrobenzene) $< \Delta S^{\neq}$ (chlorobenzene). If to exclude from this range $\Delta S^{\neq}$ (butanol), we would obtain the range: $\Delta S^{\neq}$ (styrene) $< \Delta S^{\neq}$ (dimethylformamide) $< \Delta S^{\neq}$ (nitrobenzene) $< \Delta S^{\neq}$ (chlorobenzene), which is antibate to the values of dielectric constants of the solvents: ε (dimethylformamide) $> \varepsilon$ (nitrobenzene) $> \varepsilon$ (chlorobenzene) $> \varepsilon$ (styrene).

Generalizing the results of carried out investigations it can be concluded, that the solvation of the investigated peroxides and dielectric permeability of medium are the determining factors which have an influence on kinetics of their decomposition.

The interesting are data of Table 57. The difference in values of dipole moments for transition and starting states in a case of peresters into homolysis reactions is approximately the same (0,36–0,38 D). This earnestly points to fact that the influence of polar solvent in a range of the peresters should be at least constant, and the obtained kinetic characteristics are caused by the influence of the electronic structure of compounds.

In references [112–113] we have studied in detail the associate formation peroxide–n–butyl alcohol. The results of investigations are represented in Table 58.

Despite the fact that the energy values of the hydrogen bond are insignificant, this bond can lead to the essential increasing of the peroxides decomposition rate in n–butyl alcohol leading to the considerable decreasing of a potential barrier into transition state.

It was interesting to check the correctness of the equations proposed by *Yogansen* and *Rassadin* for the systems peroxide–alcohol [114]:

$$-\Delta H = 0,5\Delta F^{0,5} \tag{62}$$

where $\Delta F^{1/2} = \Delta F^{1/2}$ (peroxide...alcohol) $- (\Delta F)^{1/2}$ («free» alcohol in carbon tetrachloride).

The values $-\Delta H$ obtained from the last equation are also included in Table 58. As we can see from this table, the complexation enthalpies $-\Delta H$ and $\Delta H'$ obtained via the different methods are coincided; this fact can be considered as an evidence of the intensities rule application possibility for the investigated systems.

Table 58. The values of hydrogen bonds energy calculated upon temperature progression of constants *(–ΔH)* and accordingly to the intensities rule *(–ΔH)*

№	Peroxy compound	T, K	$K, mole/l$	$-\Delta H,$ $kJ/mole$	$-\Delta \dot{H},$ $kJ/mole$	Composition of associate
1	 $-OOC(CH_3)_2C_2H_5$	303	$0,74 \pm 0,03$	$4,2 \pm 0,8$	$4,2 \pm 0,2$	$1:1$
		318	$0,79 \pm 0,05$			
		338	$0,88 \pm 0,05$			
2	 $-OOC(CH_3)_2C_2H_5$	303	$10^2(0,56 \pm 0,20)$	$7,1 \pm 1,2$	$7,1 \pm 1,2$	$1:2$
		318	$10^2(0,63 \pm 0,05)$			
		338	$10^2(0,75 \pm 0,07)$			
3	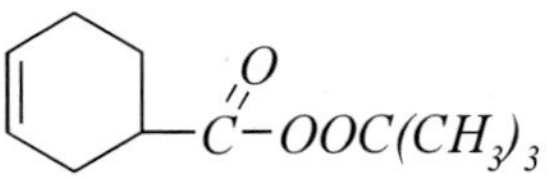 $-\overset{O}{\overset{\|}{C}}-OOC(CH_3)_3$	303	$10^4(0,23 \pm 0,01)$	$9,2 \pm 0,8$	$9,2 \pm 1,6$	$1:3$
		329	$10^4(0,28 \pm 0,02)$			
		338	$10^4(0,34 \pm 0,02)$			
4	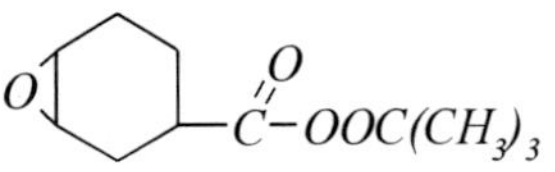 $O-\overset{O}{\overset{\|}{C}}-OOC(CH_3)_3$	303	$10^6(0,22 \pm 0,01)$	$13,4 \pm 0,8$	$13,4 \pm 0,8$	$1:4$
		318	$10^6(0,28 \pm 0,02)$			
		388	$10^6(0,36 \pm 0,03)$			
5	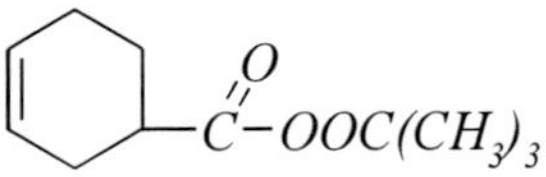 $-\overset{O}{\overset{\|}{C}}-OOC(CH_3)_3$	303	$2,00 \pm 0,04$	$3,3 \pm 0,4$	—	—
		318	$2,13 \pm 0,07$			
		338	$2,29 \pm 0,06$			
6	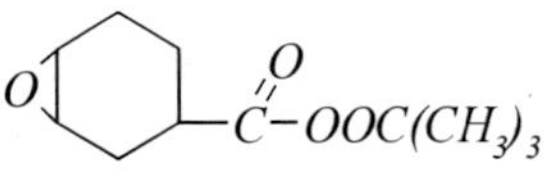 $O-\overset{O}{\overset{\|}{C}}-OOC(CH_3)_3$	303	$1,57 \pm 0,04$	$3,3 \pm 0,4$	—	—
		318	$1,68 \pm 0,04$			
		338	$1,85 \pm 0,06$			

Table 59. The values of energies of non–valence interactions in some peroxides

№	$X(O_\gamma, S)$ $R-C-OOC(CH_3)_3$ 4 5 α β	Non–valence interactions	$E_{interact.}$, a. u.
1	CH_3	$CH_3...O_\gamma$	0,1770
		$C_4...O_\alpha$	0,0576
		$(H)_3...C_5$	0,1806
			0,4152
		$O_\gamma...O_\beta$	0,0556
		$O_\gamma...O_\beta$	0,0554
		$C_5...O_\beta$	0,1460
			0,2570
2	CH_3-CH_2	$C_2H_5...O_\gamma$	0,2326
		$C_4...O_\alpha$	0,0590
		$(CH_3), (H)_2...C_5$	0,3262
			0,6178
		$O_\gamma...O_\beta$	0,0458
		$O_\gamma...O_\alpha$	0,0508
		$C_5...O_\beta$	0,1346
			0,2312
3	$(CH_3)_3C$	$(CH_3)_3C...O_\gamma$	0,3552
		$C_4...O_\alpha$	0,0554
		$3(CH_3)...C_5$	0,6410
			1,0516
		$O_\gamma...O_\beta$	0,1298
		$O_\gamma...O_\alpha$	0,0506
		$C_5...O_\beta$	0,0480
			0,2284
4	$CH_3(S)$	$CH_3...S$	0,1810
		$C_4...O_\alpha$	0,0608
		$(H)_3...C_5$	0,1576
			0,3994
		$S...O_\alpha$	0,0570
		$S...O_\beta$	0,0740
		$C_5...O_\beta$	0,1640
			0,2950

5.6. ABOUT THE FORECAST OF CHEMICAL MECHANISM AND KINETIC PARAMETERS OF PEROXIDES HOMOLYSIS

In Table 59 there are values of non–valence interactions of atoms in series of peresters which permit to do conclusions concerning the influence of the substitutes on non–valence bonds of the reactive centers of peroxides. Firstly, the stronger are non–valence interactions of substitutes R with atoms C and X of bond C–X and atoms O_α and O_β the weaker are non–valence interactions $X...O_\alpha$, $X...O_\beta$, $C...X_\beta$, Secondly, if the non–valence interaction of substitute R (Table 59) with X, O_α and O_β is not great and general stabilization energy upon non–valence bonds of reactive center is some inflated then it can be expected the synchronous homolytic decomposition of the peroxy compound.

Really, in references there are suppositions about synchronous decomposition of the $Cl-C(O_\gamma)-OOC(CH_3)_3$ peroxide. It can be seen from Table 60 that the interaction energy of Cl with $O\gamma$ is nearly in twice less than in a case of the alkyl substitutes (see Table 59). The stabilization energy upon bonds $O_\gamma...O_\alpha$, $O_\gamma...O_\beta$, $C...O_\beta$ and upon valence bonds $O_\alpha-O_\beta$, C_5-O_γ, C_5-O_α is equal to $-3,1836$ a. u., whereas for the peroxide $(CH_3)_3C-C(O)-OOC(CH_3)_3$ is equal to $-3,1699$ a. u., in other words, in the first case is some more although the comparable compounds are decomposed with approximately the same activation energies (~ 125 kJ/mole). It is follow from this that the stabilization of the reactive center of peroxide $Cl-C(O)-OOC(CH_3)_3$ is caused by some other reasons. For the considered molecule of peroxide the decrease of energy of non–valence interaction of Cl with O does not lead to increasing of the energy of non–valence interactions $O_\gamma...O_\alpha$, $O_\gamma...O_\beta$, $C...O_\beta$. Probably, the peroxide $Cl-C(O)-OOC(CH_3)_3$ decomposition proceeds in accordance with other chemical mechanism. The one among important reasons of the synchronous decomposition of peroxides is a little strength of the R–$C(O)$ bond. Really, such strength for Cl–C bond (Table 60) is considerably less than for C–C (Table 61) bond in a case of the alkyl substitutes. For the peroxides with dialkyl substitutes into activated complex the distance R–$C(O)$ is shortened that, naturally, from our considerations, leads to the increase of the non–valence interactions of substitute and attenuates in that way such interactions for the reactive center. While as the substitute is the *Cl* atom then the peroxide decomposition proceeds synchronously, that is the distance *Cl–C(O)* is increased and the interactions *Cl* are naturally decreased; this leads to the increasing of energies of non–valence bonds O_γ–O_β and especially the bond *C...O$_\beta$*. In any case these bonds into activated state for peroxide $Cl{-}\overset{\cdot\cdot O}{C}{-}OOC(CH_3)_3$ should be some stronger than in a case of alkyl substitutes since they stabilize of this state.

For the compound $F{-}\overset{\cdot\cdot O}{C}{-}OOC(CH_3)_3$ energy of *F...O$_\gamma$, F...O$_\alpha$, F...O$_\beta$* bonds interaction consists of –0,1248 *a. u.*, in other words, this value is considerably lower, than the same value for the peroxides with the dialkyl substitutes. However, the stabilization energy of reactive center $E_{O_\gamma...O_\alpha}$ + $E_{O_\gamma...O_\beta}$ + $E_{C...O_\beta}$ is higher than for the peresters with the dialkyl substitutes; especially because the strength of bond *F–C(O)* (–1,0764 *a. u.*) is considerably less than for the bond *C–C(O)* in peresters with the dialkyl substitutes. It is follows from these considerations that the synchronous homolytical decomposition of $F{-}\overset{\cdot\cdot O}{C}{-}OOC(CH_3)_3$ compound is excluded, that possible is only the homolysis of *–O–O–* bond.

For the peroxide $H_3C{-}\overset{\cdot\cdot O}{C}{-}OO{-}\overset{\cdot\cdot O}{C}{-}CH_3$ the energy of non–valence interactions *(CH$_3$...O$_\gamma$, (H)$_3$...C$_5$, C$_4$...O$_\alpha$)* is equal to –0,4342 *a. u.*, that is some more than in compound $H_3C{-}\overset{\cdot\cdot O}{C}{-}OOC(CH_3)_3$, whereas the energy of bonds *(O$_\gamma$...O$_\alpha$, O$_\gamma$...O$_\beta$, C$_5$...O$_\beta$)* is equal to –0,2543 *a. u.*, that is less, than in perester. It is follows from this, that the compound $H_3C{-}\overset{\cdot\cdot O}{C}{-}OO{-}\overset{\cdot\cdot O}{C}{-}CH_3$ homolytically should be decomposed only with the break of *–O–O–* bond.

Table 60. The values of energy of valence and non–valence interactions of atoms in molecule $Cl-\overset{,O\gamma}{\underset{5\;\;\alpha\;\;\beta}{C-OOC(CH_3)_3}}$ **into starting and transition states**

№	Bond	Starting state $-E_{1\,interact.,}$ *a. u.*	Transition state $-E_{2\,interact.,}$ *a. u.*
1	C_5-O_γ	1,3176	1,2696
2	C_5-O_α	0,9374	0,9582
3	$O_\alpha-O_\beta$	0,6838	0,1574
4	$C_5...O_\beta$	0,1294	0,0485
5	$Cl...O_\gamma$	0,0664	0,0658
		0,0146	0,0147
		0,0016	0,0013
6	$Cl...O_\alpha$	0,0438	0,0433
		0,0108	0,0091
		0,0010	0,0015
7	$Cl-C_5$	0,6138	0,6209
8	$O_\gamma...O_\beta$	0,0548	0,0306
9	$O_\gamma...O_\alpha$	0,0606	0,0759

At last, let us consider in briefly the prognosis of the reactive ability of peroxides which in their majority were not synthesized yet.

In accordance with our imaginations the peroxide compounds $R-\overset{,S}{C-OOC(CH_3)_3}$ should be thermally enough stable. In a case of $R = CH_3$, C_6H_5, *tert*–C_4H_9, CCl_3, Cl the calculated activation energies of thermolysis are equal to 174,3; 186,9; 155,4; 180,6; 136,5 *kJ/mole* respectively. The energies of $-O-O-$ bonds for these compounds are little differed from the same in peresters and are equal to –0,7498 *a. u.*, –0,7388 *a. u.*, –0,6588 *a. u.*, –0,7244 *a. u.* respectively. The stabilization energy for this series of peroxides is equal to –1,0476 *a. u.*, –1,0228 *a. u.*, –0,9854 *a. u.*, –1,0706 *a. u.*, –0,9202 *a. u.*

The values of pre–exponential factors in homolysis reactions for compounds $R-\overset{,S}{C-OOC(CH_3)_3}$, where $R = CH_3$, C_6H_5, $C(CH_3)_3$, CCl_3, Cl are respectively equal to $lgA = 18$; 19,2; 16,4; 18,5; 14,7.

Table 61. The values of energy of some valence bonds in peroxides

№	$\overset{X(O_\gamma)}{\underset{4 \quad 5 \quad \alpha \; \beta}{R{-}C{-}OOC(CH_3)_3}}$	Bond	$-E_{bond}$, a. u.
1	CH_3	$O_\alpha{-}O_\beta$	0,7466
		$C_5{-}O_\gamma$	1,2976
		$C_5{-}O_\alpha$	0,8870
		$C_4{-}C_5$	0,7560
2	C_2H_5	$O_\alpha{-}O_\beta$	0,7442
		$C_5{-}O_\gamma$	1,2842
		$C_5{-}O_\alpha$	0,9062
		$C_4{-}C_5$	0,7700
3	$(CH_3)_3C$	$O_\alpha{-}O_\beta$	0,6524
		$C_5{-}O_\gamma$	1,3136
		$C_5{-}O_\alpha$	0,9695
		$C_4{-}C_5$	0,7622
4	CH_3	$O_\alpha{-}O_\beta$	0,7498
		$C_5{=}S$	0,5708 + 0,1234 = 0,6942
		$C_5{-}O_\alpha$	0,8976
		$C_4{-}C_5$	0,7498

The peroxide $H_2C\overset{\diagup S{-}CH_3}{\diagdown OOC(CH_3)_3}$ with the stabilization energy $-1,1482$ *a. u.* should be thermally decomposed with the activation energy 207,1 *kJ/mole*, that is some more than for the analogous formal $H_2C\overset{\diagup O{-}CH_3}{\diagdown OOC(CH_3)_3}$ ($E = 175,7$ *kJ/mole*). A great value of the activation energy of compound $H_2C\overset{\diagup S{-}CH_3}{\diagdown OOC(CH_3)_3}$ homolysis is explained by a great energy of non–valence interaction $X...O_\beta$ ($-0,1752$ *a. u.*) whereas the analogous bond in $H_2C\overset{\diagup S{-}OCH_3}{\diagdown OOC(CH_3)_3}$ is only $-0,0564$ *a. u.* The strength of $-O{-}O-$ bond for the last is even some more.

For the peroxides $RCH_2OOC(CH_3)_3$ and $ClCH_2OOC(CH_3)_3$ the calculated stabilization energy is correspondingly equal to $-1,100$ *a. u.* and $-1,1666$ *a. u.*, that corresponds to the activation energies $188,3$ and $207,9$ *kJ/mole*. The strengths of $-O-O-$ bonds are equal to $-0,8086$ *a. u.* and $-0,7212$ *a. u.* More activation energy for $ClCH_2OOC(CH_3)_3$ is explained by the increasing of interaction $X...O_\beta$ $(-0,2066$ *a.* *u.* against to $-0,0954$ *a.* *u.* for $RCH_2OOC(CH_3)_3)$. Phosphorus–containing peroxide $H_2C{\overset{RH_4}{\underset{OOC(CH_3)_3}{\diagdown}}}$ with the stabilization energies $-1,1254$ *a. u.* should homolytically decomposed with the activation energy $194,6$ *kJ/mole*. The strength of $-O-O-$ bond is equal to $-0,7308$ *a. u.* In this compound enough strong is the interaction $X...O_\beta$ $(-0,1322$ *a. u.*).

The peroxide $H_3C-\overset{O}{\overset{\|}{C}}-O-\overset{CH_3}{\underset{OOC(CH_3)_3}{\overset{|}{C}}}H$ has the stabilization energy $-0,9421$ *a. u.*, that corresponds to the calculated activation energy $142,3$ *kJ/mole*. Experimental value of the activation energy of this compound decomposition in chlorobenzene is equal to $140,6$ *kJ/mole*. The investigations were carried out in a range of temperatures 403–448 *K*. This peroxide has been synthesized at the chair of synthetic rubbers of Volgograd polytechnic institute based on our recommendations. It makes a good showing via the vulcanization processes of synthetic rubbers.

Peroxides $C_6H_5OOC_6H_5$ and $C_6H_5CH_2OOC(CH_3)_3$ with the stabilization energies $-1,0004$ *a. u.* and $-0,9568$ *a. u.* should decomposed with the activation energies $144,3$ *kJ/mole* and $140,2$ *kJ/mole*.

Perester $HC{\overset{O}{\underset{OOC(CH_3)_3}{\diagup}}}$ has the stabilization energy $-1,0056$ *a. u.*, that is the calculated activation energy is near $159,0$ *kJ/mole*. Experimental value E is also $159,0$ *kJ/mole*.

However, the purposeful synthesis of the initiators needs also the solving of task on reactivity of free radicals forming from the peroxides. Let us consider some questions concerning to the prognosis of the radicals reactivity.

The primary radicals under homolysis of peroxides $R-C{\overset{S}{\underset{OOC(CH_3)_3}{\diagup}}}$ are radicals $R-C{\overset{S\bullet}{\underset{O}{\diagup}}}$. The stabilization energy of radicals $H_3C-C{\overset{S\bullet}{\underset{O}{\diagup}}}$, $C_6H_5-C{\overset{S\bullet}{\underset{O}{\diagup}}}$, $CCl_3-C{\overset{S\bullet}{\underset{O}{\diagup}}}$, $(CH_3)_3C-C{\overset{S\bullet}{\underset{O}{\diagup}}}$, $Cl-C{\overset{S\bullet}{\underset{O}{\diagup}}}$ are respectively equal to $-1,1041$ *a. u.*, $-0,0774$ *a. u.*, $-0,0829$ *a. u.*, $-0,0735$ *a. u.*, $-0,0839$ *a. u.*

Sulfur–containing radicals are more reactive than the analogous radicals $R\text{-}C{\leqslant}^{O}_{O\cdot}$. The stabilization energy of these radicals is some higher 0,11–0,16 *a. u.* The activation series of these two types of radicals are not coincided.

The values of energies of $R–C$ bonds breaking with taking into account the non–valence interactions $-C{\cdot\cdot\cdot}C{\leqslant}^{S}_{O}$ for radicals $Cl\text{-}C{\leqslant}^{S\cdot}_{O}$, $(CH_3)_3C\text{-}C{\leqslant}^{S\cdot}_{O}$, $CCl_3\text{-}C{\leqslant}^{S\cdot}_{O}$, are equal to –1,2356; –1,0398; –0,9662; –0,9595 and –0,8882 *a. u.* respectively. For radicals $R\text{-}C{\leqslant}^{O}_{O\cdot}$ these energies are some higher.

In some cases the spin density on atom with the unpaired electron can be used as the index of the reactivity. For radicals $R{-}C{\leqslant}^{S\cdot}_{O}$ the spin density is concentrated on Sulfur atom whereas in $R{-}C{\cdot}^{\cdot O}_{O}$ radicals it is distributed between two atoms of Oxygen equally. Spin density on Oxygen atoms for the considered series of radicals are in a range 0,91–0,93. However, we disposed to consider that the spin density on the atoms with the unpaired electron in alkoxyl radicals in general hardly should be by index of the reactivity, since, for example, in such complicated radicals as $H_3C{-}\overset{O}{C}\text{-}OCH(CH_3)O\cdot$, $(CH_3)_2NCH_2O^{\bullet}$ spin density is about 0,91–0,92, that is enough great; however, these radicals are absolutely un–reactive in reactions of the hydrogen atom break from any molecule.

The primary products of the peroxides $FCH_2OOC(CH_3)_3$ and $ClCH_2OOC(CH_3)_3$ homolysis are free radicals FCH_2O and $ClCH_2O$. Their reactivity advisably to compare with CH_3O. Radical $ClCH_2O$ in reactions of the hydrogen atom break from the molecule is some less reactive than CH_3O, since the stabilization energy upon bonds $H...O$ and $Cl...O$ is equal to –0,1304 *a. u.*, whereas for the CH_3O (upon bonds $H...O$) is only –0,004 *a. u.* For FCH_2O the considered characteristic is equal to –0,1146 *a. u.*, that is the radical is some more active than the $ClCH_2O$.

In decomposition reactions the F and Cl atoms should be not broken since the stabilization energy upon bonds $F–C$ and $Cl–C$ is respectively equal to –1,1297 *a. u.* and –0,7937 *a. u.*, whereas the stabilization energy needed for the hydrogen atom break from these radicals is equal to –0,3305 *a. u.* and 0,4525 *a. u.* The same magnitude for CH_3O is equal to –0,3704 *a. u.* It can be concluded from this analysis, that the most easy the hydrogen atom will be broken in radical $F–CH_2O$ with the formation of compound $F\text{-}C{\cdot}^{\cdot O}_{H}$.

As regards the isomerization reaction of these radicals, that judging upon the stabilization energy values upon C–H bonds it can be written the following range: $FCH_2O > CH_3O > ClCH_2O$. The isomerization products of these radicals are radicals $FCHOH$, CH_2OH and $ClCHOH$.

More complicated radicals $H_3C-\overset{\overset{O}{\cdot\cdot}}{C}\text{-}OCH(CH_3)\text{-}O^{\bullet}$ and $H_4P-CH_2O^{\bullet}$ are the primary products of the homolysis of $H_3C-\overset{O}{C}\text{-}O-\overset{CH_3}{CH}\text{-}OO-C(CH_3)_3$ and $H_4PCH_2OOC(CH_3)_3$. Judging upon the stabilization energy of the non–valence bonds of oxygen atom with the nearest neighbours (–0,6632 $a.\ u.$, –0,3724 $a.\ u.$, –0,40 $a.\ u.$) these complicated radicals can not be able break the hydrogen atom from any molecule. Radical $H_3C-\overset{\overset{O}{\cdot\cdot}}{C}\text{-}OCH(CH_3)\text{-}O^{\bullet}$ can be decomposed in some directions. First is the formation of $H_3C-C\overset{O}{\underset{O}{\diagup}}$ and $HC\overset{O}{\underset{CH_3}{\diagup}}{}^{\bullet}$. The second variant includes the formation of CH_3, CO_2 and $HC\overset{O}{\underset{CH_3}{\diagup}}$, the third variant forecasts the formation of $H_3C-C\overset{O}{\underset{O}{\diagup}}-C\overset{O}{\underset{CH_3}{\diagup}}$ and H, and the fourth one – the formation of CH_3 and $H_3C-C\overset{O}{\underset{O}{\diagup}}-C\overset{O}{\underset{H}{\diagup}}$.

The first variant needs the energy –1,0965 $a.\ u.$, the second one even greater energy in connection with the C–C bond break, the third variant needs the energy –0,2217 $a.\ u.$ and the forth one –0,8631 $a.\ u.$ It can be seen from these calculations, that the most energetic advantageous is the third decomposition way. We have experimentally found that really the main product of the peroxide decomposition is the compound $H_3C-C\overset{O}{\underset{O}{\diagup}}-C\overset{O}{\underset{CH_3}{\diagup}}$.

The radical H_4P-CH_2O can be decomposed probably in two ways. The first way is the formation of $H_4P-C\overset{O}{\underset{H}{\diagup}}$ and H, and the second way – with the break of P–C bond and formation of H_4P and CH_2O. In a case of the hydrogen atom detachment the stabilization energy is equal to –0,2385 $a.\ u.$ whereas for the break of P–C bond this value is a lot more –1,1444 $a.\ u.$, that is energetically advantageous is the reaction of the hydrogen atom detachment.

Summarizing the experimental data on kinetics and chemical mechanism of the peroxides it is necessary to note following.

Kinetic parameters of peroxides (peroxyacetals, peresters, nitrogen–containing peroxides) are good interpreted in accordance with the imaginations about the interaction of valence free atoms of reactive center.

Substitutes by any of the same type series in peroxides have the insignificant influence on the value of stabilization energy. The "cell" effect of solvent is displayed at the expense not only the physical–chemical properties of solvent but also at the expense of the volume and especially of the substitute conformations. Volumetric substitutes (cyclohexynil, epoxycyclohexynil) in peroxyacetals decrease the activation energy of homolysis on 8–12 *kJ/mole*.

Conformational transformations relatively to cyclohexene ring under homolysis of peroxides change the effective activation energy not more, than on 8 *kJ/mole*. For peroxy compounds with enough great stabilization energies (peroxyacetals) the influence of the specific solvation on a value of the activation energy can be neglected, since the strong intracenter interactions of atoms lead to the low effective interaction of atoms with the solvents.

In a case of the peroxides with the weak non–valence interactions of atoms in reactive center (cyclopentenyl peroxides, cyclocontaining peresters) the activation energy of homolysis at the specific solvation is decreased on 10–16 *kJ/mole*.

So, probably in the most cases an influence of the non–specific interaction of the "cell" effect, conformational transformations relatively to ring on the value of the activation energy of the peroxides homolysis can be neglected. In turn, this gives the possibilities to wide the quantum–chemical approach for estimation of the peroxides reactivity, since somehow or other it is necessary to compare the calculated characteristics of compounds with their experimental kinetic parameters.

Chapter 6

PHOTOLYSIS OF THE INVESTIGATED PEROXIDES

In this Chapter the questions concerning to kinetics and chemical mechanism of the peroxy compounds photodecomposition in solid state are considered. For the investigated compounds it is observed one *UV* absorption band with the maximum situated outside of 200 *nm*. Probably, this maximum can be referred to the absorption of $-O-O-$ group. The energy of absorbent quantum by molecules of peroxides is approximately equal to $52,3 \cdot 10^3 \ sm^{-1}$. The extinction coefficients for peroxides are situated within the interval $2–12 \cdot 10^{-4}$ at $\lambda = 210 \ nm$. The time of life into excited state τ_0 at the transition of molecule from the state $S_0 \rightarrow S_1$ for peroxides consists of $1–5 \cdot 10^{-10} \ s$. A little values of τ_0 are evidence of fact that the photodecomposition of peroxy compounds proceeds via the singlet state.

6.1. INFORMATION ABOUT PHOTODECOMPOSITION OF ORGANIC PEROXIDES

Organic peroxy compounds are easy decomposed under the *UV* illumination action as a result of a little strength of $-O-O-$ bond.

The most number of works is dedicated to the photolysis of di*tert*–butylperoxide in gaseous and liquid states [115–119]. In liquid phase the photolysis of *tert*–butyl peroxide was investigated in detail by *Bell* and *Rust* [116]. It was shown by these authors that under photolysis and thermolysis of the peroxide the same products but in some other ratio are formed.

Ethylpropyl–*tert*–butylperoxide [120], dicumylperoxide [121] and also isopropyl–*tert*–butyl peroxide [120] form the complicated alkoxyl radicals as a result of the photodecomposition $(\lambda = 253,7\ nm)$; such radicals are decomposed with the formation of methyl radicals and stable products.

Photolysis of diacyl peroxides [122] leads to the formation of radicals $R\text{-}C\overset{O}{\underset{O\cdot}{\diagdown}}$, which are easy decarboxylated. It was shown in [122] that the radical–forming process has two–quantum nature. This is, probably, explained by the adsorption of ultraviolet of peroxy and carbonyl groups.

It was shown by authors [123–125] that photo– and thermal decomposition of benzoil peroxide proceeds in singlet state.

Photolysis of the diacyl peroxides is accompanied by a great quantum yield of CO_2 [126]. This is an evidence of the high rate of intra–cell decomposition of primary oxyradicals.

The presence of esters into final products gives a basis to assume that via the photolysis process a part of the primary radicals having the exceeded energy outcome from a cell or interact with it.

Photolysis of peresters has been studied in references [127–129]. Decomposition of these compounds is not principally differed from the decomposition of diacyl peroxides.

Forming under the dialkylperoxides photolysis alkoxyradicals are decomposed with either $C\text{–}H$ or $C\text{–}C$ bonds break [130–133]. At the length of R in RO increasing the tendency of alkoxyradicals to decomposition is increased.

For the most alicyclic oxyradicals the break of cycle is energetically more advantageous than the detachment of any group [134].

6.2. PECULIARITIES OF PEROXIDES PHOTODECOMPOSITION IN SOLID PHASE

At the present time in references there is limited number of works dedicated to the photolysis in solid phase.

The hydroperoxides photolysis is considered in *ref.* [134–136]. Forming primary alkoxyradicals are highly active and easy break the hydrogen atom.

The most fully it was studied the photodecomposition of diacyl peroxides in the solid phase [137]. The all considered peroxides generate the alkyl radicals under *UV*–illumination. It was shown by authors [137] that the quantum yield determined with respect to the decomposing peroxide is equal to 1; this indicates on the one–quantum process of $-O-O-$ bond breaking.

From the point of view of the radicals pairs registration methods the most indicative are the works [138, 139], in which the radicals's pairs of nitrogen–acidulous radicals were investigated. *EPR* spectra of radicals' pairs are differed from the spectra of isolated radicals by strong dipole–dipole and exchange interaction [139].

EPR spectra in resonant field can be widened as a result of the different distance between the pairs forming radicals. From the other hand, the widening can be connected with the heterogeneousness of substance in polycrystalline and amorphous phases.

In order to understand the chemical mechanism of the radicals's pairs in solids it is necessary to study the elementary reactions of radicals in cell and outcome from the cell. Radicals's pairs can be considered as the analogue of active intermediate state for the reactions of dissociation and recombination in condensed phase.

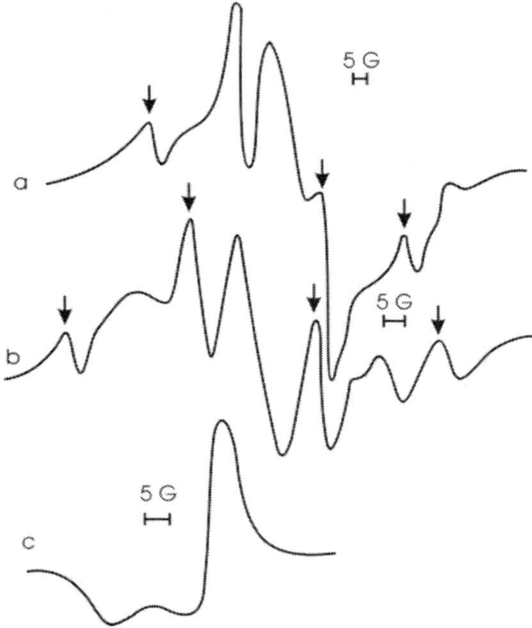

Figure 47. *EPR* spectra of free radicals forming under *UV*–illumination of *tert*–butyl peroxide at 77 K. The lines of methyl radicals are indicated by the arrows. *a* and *b* is a case of the quick and the slow cooling of *TBP* solution; *c* is the *EPR* spectrum of *UV*–illuminated peroxide in the presence of oxygen.

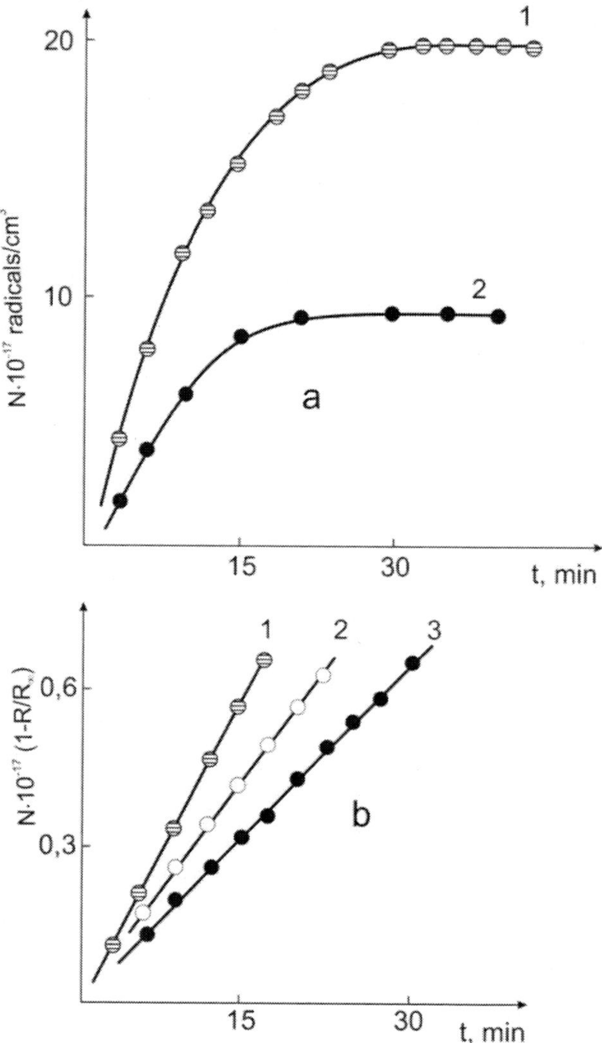

Figure 48. Kinetics of radicals accumulation and decay for *UV*–illuminated *tert*–butyl peroxide at 77 *K*. *a:* 1, 2 – increasing of the total concentration of $CH_2(CH_3)_2COOR$ and CH_3 – radicals respectively; *b:* 1, 2, 3 – accumulation of radicals $CH_2(CH_3)_2COOC(CH_3)_3$ in 30 %, 10 % and 1 % solutions of peroxide in chlorobenzene.

The base for the analysis of bimolecular reactions rates in solid phase is the *Liebiediev's* equations of formal kinetics [140] taking into account the steric factor, the number of the collisions, the number of jumps from the one cell into another and also the life time of the radical's pair.

6.3. PHOTODECOMPOSITION OF TERT–BUTYL PEROXIDE

It has been shown by us in *ref.* [141] that under photolysis of di*tert*–butyl peroxide *(TBP)* the methyl radicals are formed. Three lines with the intensities ratio 1 : 2 : 1 are appeared besides the lines characterizing for the methyl radicals under temperature increasing *(see* Figure 47*)*.

These lines were referred to the radical $CH_2(CH_3)_2COOC(CH_3)_3$. The formation of the enough great quantity of oxyisobutylene confirms of this fact. Solvents *(namely*, hexane, *tert*–butanol, chloroform and chlorobenzene) practically have not an influence on the kind of the *EPR* spectrum under the di*tert*–butyl peroxide photodecomposition.

On Figure 48 there are kinetic curves of the accumulation and decay of radicals forming under *TBP* photolysis. The limiting values of the total concentration of radicals for 1 % and 30 % of *TBP* solution in chlorobenzene are within the interval $5 \cdot 10^{16} - 10^{18}$ *spin/sm*^{3}.

Under illumination of pure peroxide the limited concentration of radicals achieves to $5 \cdot 10^{19}$ *spin/sm*^{3}. The *EPR*–spectrum in the last case is some smeared. Separation of the lines for methyl radical and for the radicals $CH_2(CH_3)_2COOC(CH_3)_3$ has been carried out by differential saturation method [142].

Experimental curves of the radicals accumulation are satisfactory described by the first order equation since in the presented case the limiting stage of the radical–chain process is the illuminated reaction (69). The rate of this reaction is near $10^{-3} \cdot 10^{-5} \approx 10^{-8}$ *mole/l·s*, whereas the rate of the illuminated reaction (65) is nearly equal to $10^{-4} \cdot 10^{-1} \, (10^{-2}) \approx 10^{-5} \div 10^{-6}$ *mole/l·s*, *i. e.* is considerably greater.

$$N = N_\infty \left(1 - e^{-k_{eff} \cdot t} \right) \tag{63}$$

For 30 % of *TBP* at 196 ^0C in chlorobenzene $K_{eff} = 12{,}5 \cdot 10^{-4} \, s^{-1}$; for 10 % of *TBP* at 196 ^0C in chlorobenzene $K_{eff} = 7{,}4 \cdot 10^{-4} \, s^{-1}$. Probably, the effective constant rate of the *TBP* photolysis some depends on concentration of peroxide.

Table 62. Products of di*tert*–butylperoxide photolysis and thermolysis in chlorobenzene (in moles per mole of decomposed peroxide)

№	Photolysis (77 *K*)		Thermolysis (398 *K*)	
	Products	*C, mole/mole*	Products	*C, mole/mole*
1	*tert*–butanol	1,06	—	0,82
2	acetone	0,58	—	0,69
3	isobutylene oxide	0,34	—	0,44
4	methane	0,57	—	0,68
5	—	—	ethane	0,03

In Table 62 there are products of the *TBP* photolysis in chlorobenzene. The concentration of peroxide was 10^{-1} *mole/l*. The time of photolysis – 4 hours.

In can be seen from the data of Table 62 that in solid and liquid phases it is formed some more *tert*–butyl alcohol than under thermolysis in liquids. This is explained by some higher reactivity of the exited *tert*–butoxyl radicals via reactions of the hydrogen atom break under photolytic decomposition of peroxides.

In solid phase the additional source of the *tert*–butanol accumulation can be the disproportionation reaction:

$$(CH_3)_3 CO + (CH_3)_3 CO \rightarrow (CH_3)_3 COH + H_2 C \overset{O}{-} C(CH_3)_2 \qquad (64)$$

Kinetic scheme of the *TBP* photolysis in solid phase can be represented by the following equations:

$$\left[(CH_3)_3 COOC(CH_3)_3 \right]^* \xrightarrow{k_1} \left[2(CH_3)_3 CO^{\cdot} \right]^* \qquad (65)$$

$$\left[(CH_3)_3 CO^{\cdot} \right]^* \xrightarrow{k_2} CH_3 + (CH_3)_2 CO \qquad (66)$$

$$CH_3 + (CH_3)_3 COOC(CH_3)_3 \xrightarrow{k_3} CH_4 + {}^{\cdot}CH_2 (CH_3)_2 COOC(CH_3)_3 \qquad (67)$$

$$\left[(CH_3)_3CO^{\cdot}\right]^* + (CH_3)_3COOC(CH_3)_3 \xrightarrow{\ k_4\ }$$

$$\rightarrow (CH_3)_3CHO + {}^{\cdot}CH_2(CH_3)_2COOC(CH_3)_3 \tag{68}$$

$$\left[{}^{\cdot}CH_2(CH_3)_2COOC(CH_3)_3\right]^* \xrightarrow{\ k_5\ } CH_3CO^{\cdot} +$$

$$H_2\overset{O}{\overset{\diagup\diagdown}{C-}}C(CH_3)_2 \tag{69}$$

$$CH_3CO^{\cdot} + CH_3CO^{\cdot} \xrightarrow{\ k_6\ } (CH_3)_3COH +$$

$$H_2\overset{O}{\overset{\diagup\diagdown}{C-}}C(CH_3)_2 \tag{70}$$

Compiling the system of the differential equations and solving it with the use of the stationary principle we will obtain the following equation:

$$[R] = [R]_\infty \left\{1 - exp(-k_5 t)\right\} \tag{71}$$

where $R = {}^{\cdot}CH_2(CH_3)_2COOC(CH_3)_3$.

From the comparison of this equation with the equation (63) we can see that $K_{eff.} = K_5$, i. e. the accumulation of $R^{\cdot}$ radicals is limited by the reaction (69).

So, the decomposition of *tert*–butyl peroxide in solid phase at 77 K proceeds in accordance with the free radical chemical mechanism. The absence of visible stationary concentration of radicals under *TBP* photolysis says about their high reactivity under photolysis conditions. The accumulation and stabilization of CH_3 radicals proceeds at the expense of their outcome from the cell. An introduction into photolyzing system of 1,1–diphenyl–2–picrylhydrizine decreases the rate of CH_3 radicals accumulation that is caused by its recombination reaction.

6.4. PHOTOLYSIS OF TERT–
ALKYLPEROXYMETHYLDIALKYLAMINES

In *ref.* [143] we have studied the photolysis in solid phase of the following compounds: *tert*–butylperoxymethyldimethylamine $(CH_3)_2NCH_2OOC(CH_3)_3$ *(I)*; *tert*–amylperoxymethyldiisopropylamine *(iso–* $C_3H_7)_2NCH_2OOC(CH_3)_2C_2H_5$ *(II)*; *tert*–amylperoxymethyldimethylamine $(CH_3)_2NCH_2OOC(CH_3)_2C_2H_5$ *(III)*; *tert*–butylperoxymethyldiethylamine $(C_2H_5)_2NCH_2OOC(CH_3)_3$ *(IV)*; *tert*–butylperoxymethylpiperidine $C_5H_{10}NCH_2OOC(CH_3)_3$ *(V)*.

Photolysis of compounds *(I)* – *(V)* was carried out both in pure medium and into solutions of chlorobenzene, diethyl ether and *n*–hexyl alcohol. *EPR*–spectra for solutions by $5 \cdot 10^{-2}$ *mole/l* concentrations are some widen that can be explained by the solvation of radicals.

On Figure 49 there are spectra of illuminated peroxides at 77 *K* representing by themselves the superposition of signals from some radicals. The attempt to separate the spectra by differential saturation method was not successful.

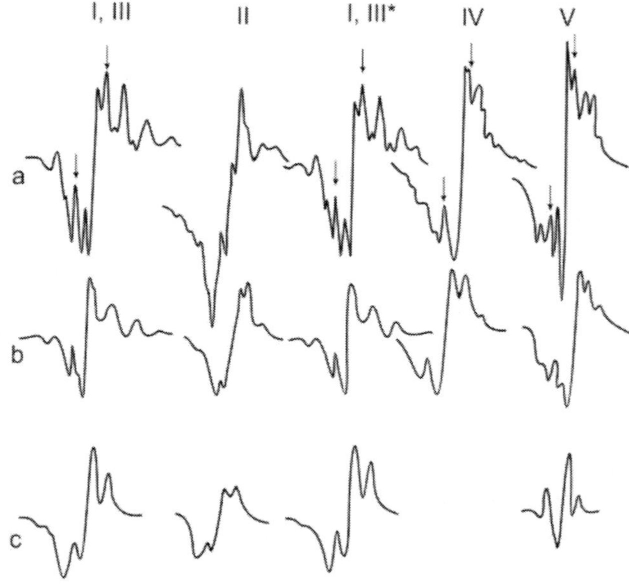

Figure 49. *EPR*–spectra of *UV*–illuminated nitrogen–containing peroxides at different temperatures: $a - 77$ *K*; $b - 93$ *K*; $c - 113$ *K*. The lines of methyl radicals are indicated by arrows. *Solutions of *I, III* in chlorobenzene; concentration 1,0 *mole/l*.

EPR–spectrum of peroxide (*I*) contains of seven lines with the constant on the nitrogen atom 33,5 *hz* and four lines with the constant 23 *hz*. The intensities ratio is approximately equal to 1 : 5 : 29 : 48 : 8 : 1 and 1 : 11 : 10 : 1 respectively. Four lines can be referred to the CH_3 radical and seven ones to the dimethylformamide one $\begin{array}{l} H_3C \\ \\ H_3C \end{array}\!\!N\!-\!C\!\cdot\!=\!O$.

In *ref.* [144–146] it was shown, that under the solid formamide illumination the analogous radicals were formed. Since the methyl groups near the atom of nitrogen of the radical $(CH_3)_2NCO$ are not equivalent [147–148], then instead of the 21 line it is observed only seven ones.

Under the temperature increasing the *EPR*–spectrum is transformed into the triplet with the constant 33,5 *hz*. Probably, in radical $(CH_3)_2NCO$ the interaction of the reactive centre with the methyl groups is weaken. The similar fact was described in *ref.* [149], in which the reaction of hydrogen atom break from the amide molecule has been studied by *EPR*–method.

The temperature increasing facilitates to the $-C=O$ group rotation around the N–C bond frequency increasing. The rotation intensity can be characterized by the correlation time τ_k or rotation frequency $\nu_k = 1/k$. In accordance with the ratio proposed in [150] it was calculated k. The rotation activation energy is equal to 10,5 *kJ/mole*.

Characteristics of *EPR*–spectra for peroxides *I–V* are represented in Table 63.

Table 63. Some parameters of *EPR*–spectra of *UV*–illuminated *tert*–alkylperoxymethyldialkylamines

Peroxide	$a_1{}^*$, ± 1,5	Number of observed lines	$a_2{}^{**}$, ± 0,5	Number of observed lines	g–factor ± 0,0002
I	33,5	7	23,0	4	2,0028
II	14,9	8	23,0	4	2,0029
III	33,5	7	23,0	4	2,0028
IV	14,9	8	—	—	2,0030
V	25,0	9	23,0	4	2,0029

Note: $a_1{}^*$ and $a_2{}^{**}$ are constants attributable to the radicals $R_1R_1NC=O$ and CH_3 respectively.

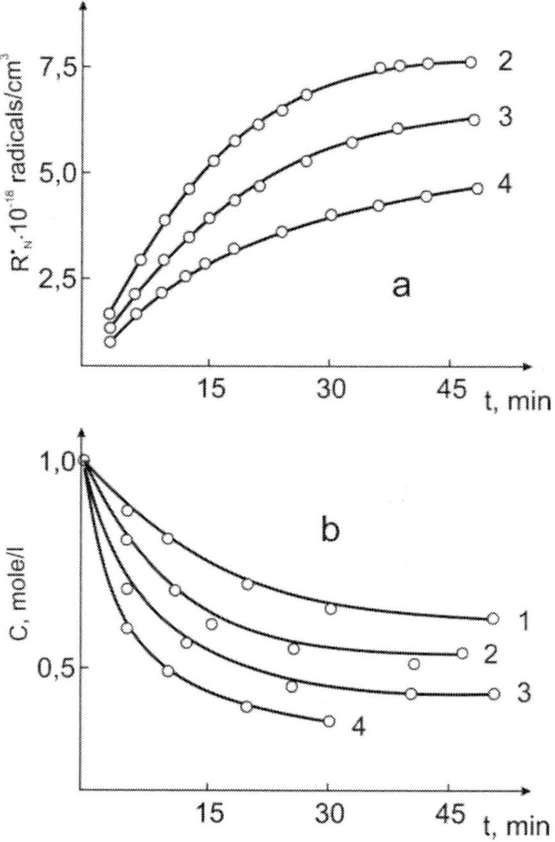

Figure 50. Kinetics of radicals $(CH_3)_2–N–C\dot{}=O$ $(R_M^{\dot{}})$ accumulation *(a)* and peroxide *I* decomposition *(b)* under its photolysis in solid phase in chlorobenzene (1 – at 77 *K*; 2 – at 93 *K*; 3 – at 103 *K* and 4 – at 113 *K*).

It is necessary to indicate that in the presence of oxygen the *EPR*–spectra have a view characterizing for peroxy radical.

6.4.1. Kinetics of Nitrogen–containing Peroxides Photolysis

Kinetics of peroxides photolysis has been studied by *EPR* method upon radical products accumulation (on relative change of the component intensity). As we can see from the Figure 50, the products accumulation rate is decreased with the temperature increasing; this is caused by their decay rate increasing. For pure peroxides *(I–V)* the value of $k_{eff.}$ at –493 *K* is equal to $1,010^{-3} – 1,210^{-}$

$^3 s^{-1}$; at $113\ K - 7,9 \cdot 10^{-4}\ s^{-1}$ are respectively equal to $2,8 \cdot 10^{-3}$, $4,0 \cdot 10^{-3}\ s^{-1}$. The activation energy is equal to $17,6\ kJ/mole$. For peroxide *IV* these constants at 98, 103 and $113\ K$ are respectively equal to $1,2 \cdot 10^{-3}$, $1,5 \cdot 10^{-3}$ and $2,8 \cdot 10^{-3}\ s^{-1}$.

Data of Table 64 permit to suppose that the chemical mechanism of the photolysis and thermolysis of peroxide *I* has a considerable likeness.

A great quantity of *tert*–butanol can be explained at the expense of the *tert*–butoxyl radical photoactivation and by a great endowment of the disproportionation reaction:

$$R_1R_1NCH_2O + OC(CH_3)_3 \rightarrow R_1R_1NC\overset{\nearrow O}{\underset{\searrow H}{}} + HOC(CH_3)_3 \qquad (72)$$

It is necessary to indicate, that with the use of special *IR* spectral data of the *UV*–illuminated peroxides (*I–V*) in chloroform and CCl_4 and in pure form at $T = 213\ K$ it was shown that the products of photolysis are accumulated in solid phase.

Starting from the data of kinetics of radicals accumulation and decay, identification of the forming radicals and products of photolysis, the decomposition of nitrogen–containing peroxides can be represented by the following scheme:

$$R_1R_1NCH_2O_tAlk \xrightarrow{\ h\nu\ } [R_1R_1NCH_2O_tAlk]^* \rightarrow [R_1R_1NCH_2O]^* + [O_tAlk]^* \qquad (73)$$

Table 64. Products of photolysis and thermolysis of *tert*–butylperoxymethyldimethylamine

№	Photolysis (77 K)		Thermolysis (401K)	
	Products	$C,$ *mole/mole*	Products	$C,$ *mole/mole*
1	*tert*–butanol	0,90	—	0,66
2	acetone	0,10	—	0,32
3	isobutylene oxide	0,03	—	—
4	dimethylformamide	0,99	—	0,91
5	methane	0,10	—	0,22
6	hydrogen (H_2)	0,05	—	0,05
7	—	—	ethane	0,05

$$[R_1R_1NCH_2O]^* \to H + R_1R_1NC{\overset{O}{\underset{H}{\lessgtr}}} \tag{74}$$

$$[O_tAlk]^* \to R^\bullet + \textit{inactive product} \tag{75}$$

$$[R_1R_1NCH_2O]^* + [O_tAlk]^* + M \to R_1R_1NC{\overset{O}{\underset{H}{\lessgtr}}} + HO_tAlk \tag{76}$$

$$[O_tAlk]^*, [R, H] + R_1R_1NCH_2O_tAlk \to$$
$$\to [HO_tAlk, RH + H_2] + R_1R_1NC{\overset{O}{\underset{H}{\lessgtr}}} + {}^\bullet O_tAlk \tag{77}$$

$$[HO_tAlk, R^\bullet, H] + R_1R_1NC{\overset{O}{\underset{H}{\lessgtr}}} \to [HO_tAlk, RH, H_2] + R_1R_1NC^\bullet{=}O \tag{78}$$

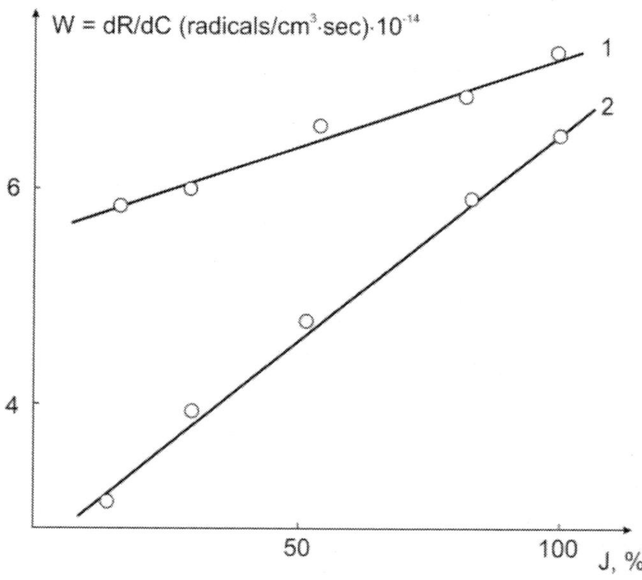

Figure 51 *a*. Dependence of radical–forming rate on intensity of *UV*–illumination for 9 (2) and 33 (1) minutes for peroxide *I* after illumination staring respectively.

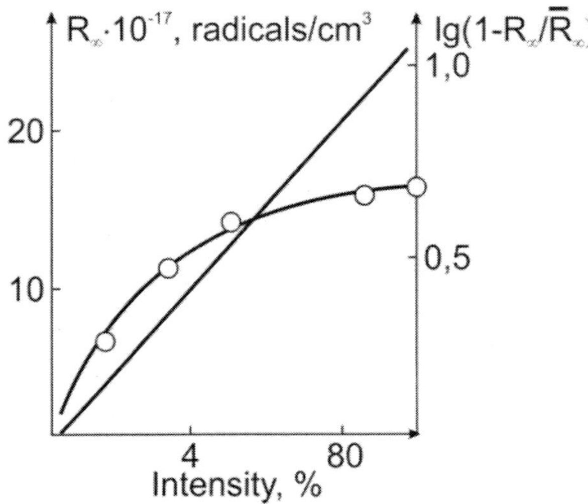

Figure 51 *b*. Dependence of limiting concentration of radicals by formamide type for peroxide *I* at 77 *K* on intensity of *UV*–illumination and its linear anamorphosis.

Maximal concentration of radicals *(CH₃)₂NCO* at 77 *K* not exceeded of $8 \cdot 10^{18}$ *spin/sm*^{3}. The value of $K_{eff.}$ obtained at 77 *K* for peroxides *I*, *IV* and *V* are approximately equal to $6 \cdot 10^{-3}$ *s*$^{-1}$. The magnitude of $K_{eff.}$ founded upon the ˙*CH₃* radicals accumulation rate (peroxide *I*) in hexyl alcohol is equal to $13,4 \cdot 10^{-4}$ *s*$^{-1}$, whereas in chlorobenzene this value is equal to $6 \cdot 10^{-4}$ *s*$^{-1}$.

The rate of radicals *(CH₃)₂NCO* accumulation depends on illumination intensity *(Figure 51 a, b)*.

As we can see from Figure 51 the dependence of R_∞ on intensity is kept only to 30 % of value of maximal intensity of illumination source mercury lamp *DRSh*–1000.

Curve *δ* on Figure 51 is satisfactory described by equation $R = R_\infty(1-exp(-\gamma I))$, where R_∞ is limiting concentration at *UV* illumination by full light and intensity 100 %. Presented equation in semi logarithmic coordinates is represented by the straight line. Coefficient *γ* for peroxide *I* is equal to 1,1. At low illumination intensities $exp(-\gamma I) = 1-\gamma I$, i. e. $R = R_\infty I$. In these cases the linear dependence of R_∞ on *I* should be observed. This is in good agreement with the obtained experimental values of R_∞ under illumination of lamp with the maximal power 275 *W*.

6.4.2. Kinetics of Radicals' Decay

The kinetic curve of decay of radicals which are formed under photolysis of the peroxides *I* and *V* is represented on Figure 52.

Authors of [151–152] assume that the gradation of radicals decay process on temperature is connected with the local phaseous transformations in solid matrix.

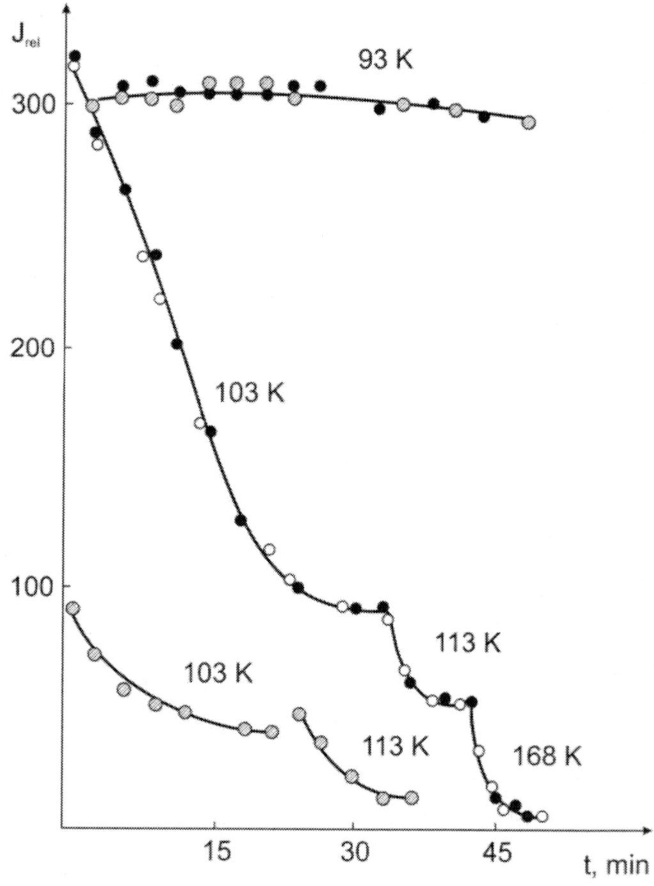

Figure 52. Kinetics of decay of radicals by $\overset{R\diagdown}{\underset{R\diagup}{N}}-\overset{\cdot}{\underset{O}{C}}$ type under different temperatures; (o, •, ⊗ are radicals of the illuminated peroxides *I*, *IV* and *V* respectively).

Kinetics of decay of $R_1R_1NC\dot{O}$ radicals at 77 K and higher is described by the first order equation. In a case of temperatures 103 K, 113 K and 123 K the rate constants of radicals decay for peroxide I

$$R_1R_1N\overset{\cdot}{C}O + \begin{cases} R_1R_1NCH_2OO_tAlk \\ R_1R_1NCHO \end{cases} \rightarrow \begin{cases} R_1R_1NCHO + O_tAlk + R_1R_1NCHO \\ R_1R_1NCHO + R_1R_1NCHO \end{cases}$$

$$(79)$$

Judging on the photolysis products of the nitrogen–containing peroxides via reactions of R, H with the peroxides and dialcylformamides, the stages (77) and (78) can be neglected in whole. More detailed consideration of scheme is given in Chapter 6.8.

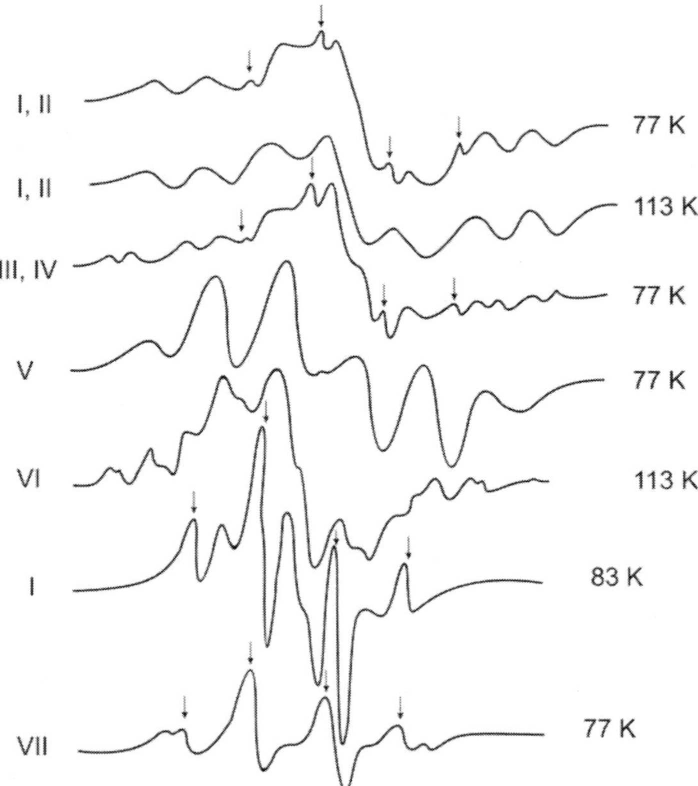

Figure 53. *EPR* spectra of radicals forming under photolysis of the individual *tert–*butylperoxyacetals in chlorobenzene at 77 K. The lines of the methyl radicals indicated by the arrows.

6.5. PHOTODECOMPOSITION OF TERT.–BUTYLPEROXYACETALS

It has been investigated the following peroxyacetals: methoxy*tert.*–butylperoxycyclohexenylmethane (*I*), *n*–propoxy*tert*–butylperoxycyclohexenyl-methane (*II*), methoxy*tert*–butylperoxy(3,4–epoxyhexyl)–methane (*III*), ethoxy*tert*–butylperoxy(3,4–epoxycyclohexyl)methane (*IV*), methoxy*tert*–butylperoxyhexylme-thane (*V*), methoxy*tert*–butylperoxyphenylmethane (*VI*), methoxytert–butylperoxy-methane (*VII*).

On Figure 53 there are *EPR* spectra of radicals forming under the photolysis of peroxides. Four lines indicated on Figure 54 by arrows belong to the ˙*CH₃* radical with the constant $23,0 \pm 0,5$ *hz*.

In Table 65 there are parameters of *EPR*–spectra of observed radicals.

The constants in cyclohexenyl radical have been determined only for protons lying in positions 1 and 6.

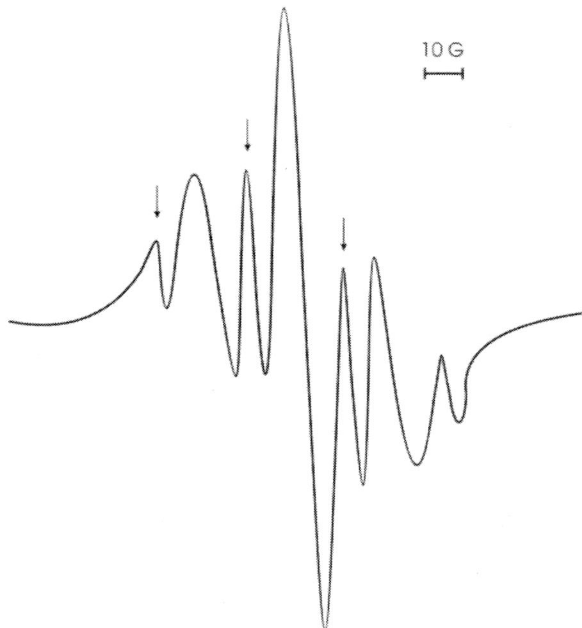

Figure 54. EPR spectrum of radicals forming under photolysis of the peroxide (*I*) at 77 K. The lines of the methyl radicals indicated by the arrows.

Table 65. Constants of radicals forming under the *UV*–illumination of peroxyacetals

Peroxide	Radicals	*g*–factor ± 0,0002	Constants, *hz*
I	(cyclohexenyl radical structure, positions 4, 2, 6, $C\cdot$, CH_3)	*2,0031*	$a_{CH3} = 23$
			$a_{H6} = 43,7$
			$a_{H1} = 17,5$
III	(3,4-epoxycyclohexyl radical structure, O, positions 4, 2, 6, $C\cdot$, CH_3)	*2,0031*	$a_{CH3} = 23$
			$a_{H6} = 48$
			$a_{H1} = 27,4$
			$a_{H5} = 11,4$
V	$H_3C-(CH_2)_4-\overset{H}{\underset{H}{C}}\cdot$	*2,0031*	$a_{H\alpha} = a_{H\beta 1} = 1/2 a_{H\beta 2} = 22$
VI	(phenyl-substituted radical structure, positions 4, 2, 6, $C\cdot$, CH_3)	*2,0031*	$a_{CH3} = 23$
			$a_{H2,6} = 18,9$
			$a_{H3,5} = 6,1$
			$a_{H4} = 2,2$
VII	*has been not identified*	*2,0028*	$a_{CH3} = 23$

An appearance of the six–linear *EPR* spectrum is an evidence of the deferred inversion of the cyclohexenyl radical in solid matrix at 77 *K* and the anisotropic hyperfine interaction *(HFI)* with the protons 2, 3, 4 are not averaged. That is why the width of the individual lines exceeds the *HFI* constants in these positions and the number of spectrum components instead of the expecting 18 lines is decreased to the 6 ones. Analogous fact of the high frequency of inversion at low temperatures is noted at studies of the cyclohexane radiolysis in *ref.* [153]. An appearance of *HFI* for two protons in position 6, probably is caused by near location of double bond. In radical, the molecular orbital, including the double bond, overlaps the sixth atom of carbon. In a case of 3,4–epoxycyclohexyl radical on the residuary proton the *HFI* constant is increased in 1,5 time ($a_1 = 27,5$ *hz*) and slightly increases the splitting from two protons near sixth atom of carbon ($a_{H6} = 48$ *hz*). The view of the spectrum for this radical is not changed under temperature increasing till 123 K.

The *EPR* spectrum of the phenyl radical represents by itself the triplet with the splitting 18,9 *hz* for *ortho*–substituted hydrogens and little splitting for *meta*–substituted atoms of hydrogen.

An increase of the temperature strongly decreases the concentration of phenyl radicals. An interpretation of spectra is strongly complicated due to the ratio *signal/ noise*. At temperatures above 127 K it was for sure possibly to evolve only the central line of the *EPR* spectrum.

We have done the theoretical calculation of the phenyl radical under approximation. Results of calculations are represented in Table 66. The calculated values of *HFI* constants are satisfactory agreed with their experimental values.

Table 66. Spin densities and *HFI* constants of phenyl radical

	a^H exper. (hz) $\pm$ 1,0	a^H calc. [**] (hz)	ρ_Σ calcul.	ρ_S calcul. [**]
C_1	140[*]	150[*]	0,9056	0,1835
C_2	—	—	−0,0555	−0,0058
C_3	—	—	0,0747	0,0131
C_4	—	—	−0,0459	−0,0032
C_5	—	—	0,0747	0,0131
C_6	—	—	−0,0555	−0,0058
HC_2	18,9	19,245	0,0356	0,0356
HC_3	6,1	6,270	0,0116	0,0116
HC_4	2,2	3,949	0,0073	0,0073
HC_5	6,1	6,270	0,0116	0,0116
HC_6	18,9	19,245	0,0356	0,0356

Note:

ρ_Σ is total spin density.

ρ_S is orbital endowment into spin density.

a^H is HFI constant.

[*] is calculated and experimental HFI constant for core C_1^{13} [30].

[**] are calculations, which have been done by Turovsky M. A.

Six lines with the ratio of relative intensities of *EPR*–spectrum 1 : 3 : 4 : 4 : 3 : 1 belong to the $n–C_6H_{13}$ radical. Hyperfine structure of this radical is caused by the interactions of unpaired electron with α and β–protons situated near the reactive center.

6.5.1. Kinetics of Radicals' Accumulation

On Figure 55 there is typical curve of radicals' accumulation under photolysis of peroxyacetals. The value of limiting concentration of radicals under peroxides illumination in pure form not exceeds $10^{18} - 10^{19}$ *spin/sm³* (at 77 *K*). This value is decreased with the temperature increasing. For example, for peroxides *I, II, III* at 103 *K* $R_\infty = 1,0 \cdot 10^{17}$ *spin/sm³*, at 113 *K* $P = 5 \cdot 10^{16}$ *spin/sm³*.

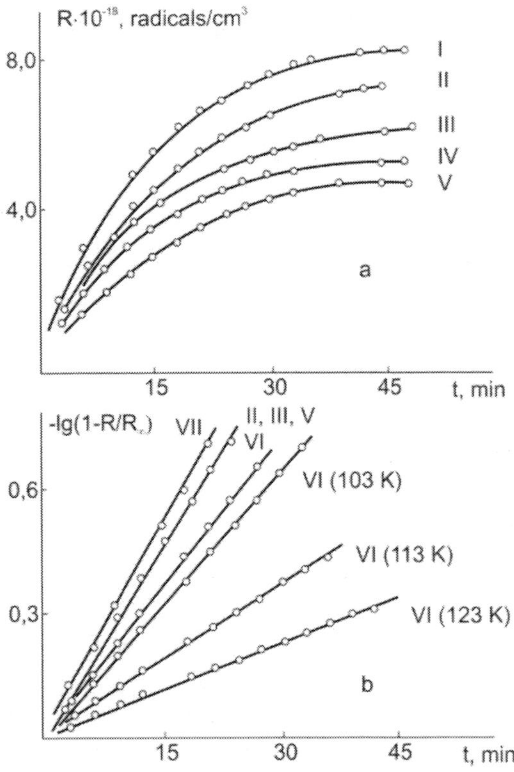

Figure 55. Kinetic curves of radicals *(a)* under photolysis of *tert*–butylperoxyacetals in pure form (77 *K*) and some their semi–logarithmic anamorphosises *(b)*.

Table 67. Products of the photolysis and thermolysis of methoxy*tert*–butylperoxycyclohexenylmethane (I) and methoxy*tert*–butylperoxyphenylmethane (II)

№	Photolysis (77 K)		Thermolysis (413 K)	
	Products	$C,\ mole/mole$	Products	$C,\ mole/mole$
(*I*)				
1	*tert*–butanol	0,83	—	0,40
2	acetone	0,15	—	0,55
3	isobutylene oxide	0,03	—	—
4	$H-\overset{O}{\underset{}{C}}-OR_2$	0,37	—	0,19
5	$R_1-\overset{O}{\underset{}{C}}-OR_2$	0,62	—	*it wasn't determined*
6	methane	0,14	—	0,55
7	R_1H	0,36	—	0,19
8	H_2	0,06	—	—
9	—	—	ethane	0,03
(*II*)				
1	*tert*–butanol	0,51	—	0,35
2	acetone	0,43	—	0,63
3	$H-\overset{O}{\underset{}{C}}-OR_2$	0,33	—	0,19
4	$R_1-\overset{O}{\underset{}{C}}-OOR_2$	0,66	—	*it wasn't determined*
5	R_1H	0,34	—	0,19
6	methane	0,42	—	0,55
7	H_2	0,05	—	—
8	—	—	ethane	0,02

The value of $K_{eff.}$ at 77 K for peroxides *I, III, IV, V* is approximately the same and is equal to $11{,}0 \cdot 10^{-4}\ s^{-1}$. An increase of the temperature leads to the $K_{eff.}$ decreasing. For example, for peroxide *VII* $K_{eff.} = 8{,}6 \cdot 10^{-4}\ s^{-1}$ (103 K), $K_{eff.} = 4{,}6 \cdot 10^{-4}\ s^{-1}$ (113 K), $K_{eff.} = 1{,}9 \cdot 10^{-4}\ s^{-1}$ (123 K). Data concerning to the final products of photolysis in solid phase for some peracetals are represented in Table 67.

Judging upon data of the chromatographic analysis of methoxy–*tert*–butylperoxycyclohexanylmethane (*I*) and methoxy–*tert*–butylperoxyphenylmethane (*II*) photolysis (*see* Table 67) it is necessary to suppose about a great endowment of the disproportionation reaction in the cell with the formation of compound $H-\overset{\overset{O}{\|}}{C}-OR_2$, that is

$$[R_1HC(OR_2)O + OC(CH_3)_3] \rightarrow R_1COOR_2 + HOC(CH_3)_3 \tag{80}$$

The compound $H-\overset{\overset{O}{\|}}{C}-OR_2$ can be also formed in accordance with the reaction

$$\begin{matrix} R_1\diagdown \\ \diagup C\diagdown \\ H \quad O\bullet \end{matrix}\overset{OR_2}{\diagup} \longrightarrow R_1C\overset{OR_2}{\underset{O}{\diagdown}} + H \tag{81}$$

However, judging upon the products, the last way is less probable.

Product of the $H-\overset{\overset{O}{\|}}{C}-OR_2$ photolysis is a result of the reaction

$$\begin{matrix} R_1\diagdown \\ \diagup C\diagdown \\ H \quad O\bullet \end{matrix}\overset{OR_2}{\diagup} \longrightarrow R_1^{\bullet} + HC\overset{OR_2}{\underset{O}{\diagdown}} \tag{82}$$

This compound is formed in two times less at the thermolysis, than at the photolysis (*see* Table 67). Under the same conditions of the thermolysis it is always formed less of *tert*–butanol and more acetone; that is the disproportionation reaction is always less probable. Formation of the exceeded quantity of *tert*–butanol under the photolysis conditions can be explained by the reaction of *tert*–butoxyl activated radical with the solvent.

Table 68. Parameters of *EPR* spectra of the *UV*–illuminated peresters

№	Radicals	HFI Constants, hz $\varDelta\alpha \pm 1{,}0$	The number of lines	Spin density on atom of carbon
I	C_2H_5	$a_\alpha^H = 22$; $a_\beta^H = 27$	12	0,944
II	$n{-}C_3H_7$	$a_\alpha^H = 22$		
III	$n{-}C_4H_9$	$a_\beta^{H1} = \dfrac{1}{2}$; $a_\beta^{H2} = 22$	6	0,944
IV	$n{-}C_6H_{13}$			
V	(cyclohexadienyl radical, CH_3)	$a_{CH3}^H = 23$		
		$a_{2,6}^H = 18{,}9$		
		$a_{3,5}^H = 6{,}1$		
		$a_4^H = 2{,}2$	6	1,0
VI	(O-substituted cyclohexadienyl radical, CH_3)	$a_{CH3}^H = 23$		
		$a_6^H = 48$		
		$a_1^H = 27{,}4$		
		$a_5^H = 11{,}4$	10	—

Table 69. Products of some esters photolysis and thermolysis

$$R_1\overset{\overset{O}{\|}}{C}-OOC(CH_3)_3 \quad (R_1 = C_6H_5 \text{ (I); } C_2H_5 \text{ (II); } n{-}C_3H_7 \text{ (III)})$$

№	Photolysis (77 K)		Thermolysis (409 K)	
	Products	C, mole/mole	Products	C, mole/mole
(I)				
1	*tert*–butanol	0,80	—	0,42
2	acetone	0,20	—	0,56
3	methane	0,24	—	0,31
4	CO_2	0,98	—	0,99
5	—	—	ethane	0,12
(II)				
1	*tert*–butanol	0,75	—	0,42
2	acetone	0,24	—	0,56
3	methane	0,24	—	0,35
4	CO_2	0,98	—	0,98
5	*n*–hydrocarbons	0,99	—	0,98
6	—	—	ethane	0,10
(III)				
1	*tert*–butanol	0,76	—	0,56
2	acetone	0,23	—	0,38
3	methane	0,23	—	0,18
4	CO_2	0,98	—	0,99
5	*n*–hydrocarbons	0,98	—	0,98
6	—	—	ethane	0,10

6.6. PERESTERS PHOTODECOMPOSITION

Photolysis of peresters has been studied at 77 K in a pure form and into solutions of hexane. The compounds by general structure $R\overset{\overset{O}{\|}}{C}-OOC(CH_3)_3$, where $R = C_2H_5$ (*I*), $n{-}C_3H_7$ (*II*), $n{-}C_4H_9$ (*III*), $n{-}C_6H_{13}$ (*IV*), ⬡ (*V*), o⬡ (*VI*) have been used as the objects of the investigations.

The main parameters of *EPR* spectra are represented in Table 68.

As we can see from the data of Table 68, for radicals $n–C_3H_7$, $n–C_4H_9$, $n–C_6H_{13}$ the *HFI* constants are the same; this means, that the free electron interacts only with the near surroundings.

HFI constants for interaction of electron with the protons into β–position and with the one from the protons into α–position are equal to 22 *hz*. The second proton into β–position generates an additional splitting of the *EPR* spectrum equal to 44 *hz*. The value of limiting concentration of the accumulating radicals under peresters' photolysis in a pure form and in hexane not exceeds the 10^{19} $spin/sm^3$. The effective rates constants of the radicals' accumulation for the all studied esters are within the interval $9–11 \cdot 10^{-4}$ s^{-1} [154].

In Table 69 there are products of the peresters' photolysis and thermolysis into hexane.

It can be seen from the Table 69, that a lot more *tert*–butanol and a less quantity of acetone are formed at the photolysis. It is simply explained by a higher reactivity of the activated *tert*–butoxyl radicals. Oxyacyl radicals, probably, are easy decomposed under the action of ultraviolet with the formation of carbonic gas and alkyl radicals which excavate the hydrogen atom from the solvent with the formation of hydrocarbons.

6.7. PHOTOLYSIS OF SILICON–CONTAINING PEROXIDES

It has been investigated the following silicon–containing compounds:

$(CH_3)_3SiOOC(CH_3)_3$	(I);
$(CH_3)_3COOSi(CH_3)_2OOC(CH_3)_3$	(II);
$CH_3O(CH_3)_2SiOOC(CH_3)_3$	(III).

EPR spectrum of radicals under illumination of pure peroxides is practically not differed from *EPR* spectra into chlorobenzene. However, the width of spectra for the pure peroxides is some less than for the solutions. On Figure 54 *(see above)* there is *EPR* spectrum of the *UV*–illuminated peroxide (I). As it can be seen from this figure, *EPR* spectra represent by themselves the superposition consisting of two spectra. The lines corresponding to the methyl radicals are pointed by the arrows. The *HFI* constant for the fourth lines of the methyl radical consists of 23,0 *hz*. Besides of these fourth lines three lines are clearly observed with the approximate ratio 1 : 2 : 1 which can be respect to the CH_2–fragments of the itself peroxide. An appearance of such type of radicals is probably caused by the hydrogen atom break from the molecule of peroxide.

The *HFI* constant for these two atoms of hydrogen consists of 23,0 *hz*. Hyperfine splitting of the unpaired electron with others protons is evidently so less that it is not observed due to anisotropic factors. At the temperature increasing to 93 *K* the lines of the methyl radicals are disappeared; only three lines corresponding to CH_2–fragments are remained.

An absence of the lines corresponding to oxytrimethylsilyl radical in *EPR* spectrum can be explained by its high reactivity [155]. K_{eff} for the presented peroxides at 77 *K* are equal to $1,0 \cdot 10^{-3}$ s^{-1}; $0,9 \cdot 10^{-3}$ s^{-1}; $1,22 \cdot 10^{-3}$ s^{-1} respectively.

Analysis of the peroxides photodecomposition products is represented in Table 70.

On the basis of *EPR* data and chromatographic analysis it can be proposed the following scheme of the investigated compounds photodecomposition:

$$\left[(CH_3)_3 SiOOC(CH_3)_3\right]^* \xrightarrow{k_1} \left[(CH_3)_3 SiO\right]^* + \left[(CH_3)_3 CO\right]^* \quad (83)$$

Table 70. Final products of the photolytic and thermal decomposition of the silicon–containing peroxides (mole per mole of the decomposed peroxide)

Products	Peroxide (*I*)		Peroxide (*II*)	
	Photolysis	Thermolysis	Photolysis	Thermolysis
acetone	0,15	0,50	0,14	1,40
tert–butanol	0,64	0,26	1,70	0,51
hexamethyl-disiloxane	0,13	0,42	—	—
silanol	0,68	0,16	—	—
isobutylene oxide	0,11	—	0,08	—
methane	0,14	0,48	0,10	1,12

$$\left[(CH_3)_3 C\overset{\cdot}{O}\right]^* \xrightarrow{k_2} (CH_3)_2 CO + CH_3 \qquad (84)$$

$$\left[(CH_3)_3 C\dot{O}\right]^* + \left.\begin{array}{l}(CH_3)_3\,SiOOC(CH_3)_3 \\ (CH_3)_3\,COOSi(CH_3)_3\end{array}\right\} \xrightarrow{\;k_3,k_3'\;}$$

$$\rightarrow (CH_3)_3\,COH + \left\{\begin{array}{l}CH_2(CH_3)_2\,SiOOC(CH_3)_3 \\ (CH_3)_3\,SiOOC(CH_3)_2\,CH_2\end{array}\right. \tag{85}$$

$$\left[(CH_3)_3 Si\dot{O}\right]^* + \left.\begin{array}{l}(CH_3)_3\,SiOOC(CH_3)_3 \\ (CH_3)_3\,SiOOC(CH_3)_3\end{array}\right\} \xrightarrow{\;k_4,k_4'\;}$$

$$\rightarrow (CH_3)_3\,SiOH + \left.\begin{array}{l}CH_2(CH_3)_2\,SiOOC(CH_3)_3 \\ (CH_3)_3\,SiOOC(CH_3)_2\,CH_2\end{array}\right\} \tag{86}$$

$$CH_3 + (CH_3)_3\,SiOOC(CH_3)_3 \xrightarrow{\;k_5\;} CH_4 + CH_2(CH_3)_2\,SiOOC(CH_3)_3 \tag{87}$$

$$CH_3 + (CH_3)_3\,COOSi(CH_3)_3 \xrightarrow{\;k_6\;} CH_4 + CH_2(CH_3)_2\,COOSi(CH_3)_3 \tag{88}$$

$$\left[CH_2(CH_3)_2\,SiOOC(CH_3)_3\right]^* \xrightarrow{\;k_7\;} (CH_3)_2\,SiCH_2O + (CH_3)_3\,COH \tag{89}$$

$$\left[CH_2(CH_3)_2\,SiOOC(CH_3)_3\right]^* \xrightarrow{\;k_8\;} (CH_3)_3\,SiOH +$$

$$(CH_3)_2 C{\underset{CH_2}{\overset{O}{\diagup\!\!\diagdown}}} \tag{90}$$

$$(CH_3)_3\,CO + (CH_3)_3\,CO \xrightarrow{\;k_9\;} (CH_3)_3\,COH + \;(CH_3)_2 C{\underset{CH_2}{\overset{O}{\diagup\!\!\diagdown}}} \tag{91}$$

$$(CH_3)_3\,SiO + (CH_3)_3\,CO \xrightarrow{\;k_{10}\;} (CH_3)_3\,SiOH + \;(CH_3)_2 C{\underset{CH_2}{\overset{O}{\diagup\!\!\diagdown}}} \tag{92}$$

$$(CH_3)_3\,SiO + (CH_3)_3\,SiO \xrightarrow{\ k_{11}\ } (CH_3)_3\,SiOH +$$

$$(CH_3)_2Si\diagup_{\substack{O \\ | \\ CH_2}}$$

$$\tag{93}$$

$$2(CH_3)_3\,SiOH \rightarrow (CH_3)_3\,SiOSi(CH_3)_3 \tag{94}$$

Let us indicate $(CH_3)_3SiOOC(CH_3)_3$ as (P); $[(CH_3)_3CO]$ as (OTB); $[(CH_3)_3SiO]$ as $(OTSi)$.

Applying the stationarity principle and assuming that $k_3 = k_3'$, $k_4 = k_4'$, $k_5 = k_6$, $k_7 = k_8$, $k_9 = k_{10} = k_{11}$, $\dfrac{dR_1}{dt} = \dfrac{dR_2}{dt}$ we will obtain the expression for the total rate:

$$\frac{dR}{dt} = \frac{d[R_1]}{dt} + \frac{d[R_2]}{dt} = \frac{d[R_1 + R_2]}{dt} = 2\frac{d[R_1]}{dt}$$

$$\frac{d[R_1]}{dt} = [P] \cdot \frac{k_1 \cdot k_2 + 2k_3\sqrt{\dfrac{2k_1[P]}{k_9}}(k_2 + k_3 \cdot P) + 4k_1 k_3 P}{2\left(k_2 + 2k_3 P + \sqrt{2k_1 \cdot k_9 P}\right)} -$$

$$-\frac{\left(k_2 + 4k_3 P + 2\sqrt{2k_1 \cdot k_9 P} \cdot k_7\right) \cdot k_7}{2\left(k_2 + 2k_3 P + \sqrt{2 \cdot k_1 \cdot k_9 P}\right)} \cdot R_1 \tag{95}$$

This equation can be written in a form

$$\frac{d[R_1]}{dt} = A - B[R_1]$$

The solution of this equation is the expression $[R_1] = \left[\dfrac{A}{B}\right]\left(1 - e^{-Bt}\right)$

$$[R_1] = [R_1]_\infty \left(1 - e^{-Bt}\right)$$

$$[R] = 2[R_1] = 2[R_1]_\infty \left(1 - e_{-Bt}\right)$$

$$[R] = [R]_\infty \left(1 - e^{-Bt}\right)$$

$$[R]_\infty = 2\frac{A}{B} = 2[P] \cdot \frac{k_1 \cdot k_2 + 2k_3 \sqrt{\dfrac{2k_1 P}{k_9}\left(k_2 + 2k_3 P\right) + 4k_1 \cdot k_3 [P]}}{\left(k_2 + 4k_3 P + 2\sqrt{2k_1 \cdot k_9 P}\right) \cdot k_7}$$

$$k_{eff.} = B = \frac{\left(k_2 + 4k_3 P + 2\sqrt{2k_1 \cdot k_9 P}\right) \cdot k_7}{2\left(k_2 + 2k_3 P + \sqrt{2k_1 \cdot k_9 \cdot P}\right)}$$

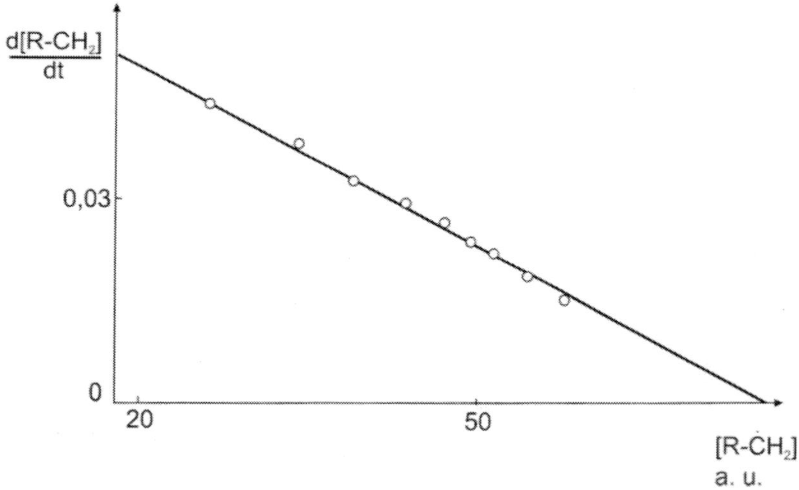

Figure 56. Dependence of the radicals with CH_2–fragment accumulation rate on their concentration for peroxide (*I*) at 77 *K*.

As we can see, $k_{eff.}$ is enough complicated. However, the radicals $CH_2(CH_3)_2SiOOC(CH_3)_3$ and $(CH_3)_3SiOOC(CH_3)_2CH_2$ accumulation rate via time is described by the first order law reaction *(see* Figure 56*)*, *i. e.* the experimental equation $[R] = [R]_\infty \left(1 - e^{-kt}\right)$ is formally agreed with the theoretically founded one $[R] = [R]_\infty \left(1 - e^{-k_{eff.}t}\right)$.

Quantum yields of the investigated peroxides in liquid phase were measured in a wide range of concentrations from $1 \cdot 10^{-3}$ till 0,1 *mole/l*. In a range of the concentrations from $1 \cdot 10^{-3}$ till $2 \cdot 10^{-2}$ *mole/l* the quantum yields of peroxides don't depend on the starting concentrations and are near to 1 [130].

At higher starting concentrations the quantum yield of a process is increased; this is an evidence of the induced peroxides decomposition. Since only the quant absorption is necessary for the initiation of the primary process, then the rates of the primary processes are in direct proportion to the intensity of the absorbed quant [130].

6.8. ABOUT «HOT» MOLECULES AND RADICALS AT THE PEROXIDES PHOTODECOMPOSITION

Mc. Millan [156] was observed the *«hot effect»* under photolysis of the dialkylperoxides. The peroxides photodecomposition reaction rate in the presence of the radicals' acceptor has been registered only to the limited magnitude. It was done the conclusion by the above–said author that a part of the alkoxyradicals into reactive mix is in the activated state and undergoes to the accelerated decomposition. Under the photolysis conditions ($\lambda = 2537 \ \overset{0}{A}$) the energy excess consisting of 293–314 *kJ/mole* for the studied peroxides is distributed between two alkoxyradicals.

In accordance with the *Tyudesh* [157] the effect of «hot radicals» is taken into account with the use of some probability factor θ, which is determined by a part of «hot radicals» able to decompose in the presence of inhibitor. The value θ considerably depends on energy of the absorbed quantum by the compound.

In liquid and solid phases the response of energy by the activated particle proceeds via its cooperative interaction with the neighbours.

It is generally well–known, that the dissociation energy of $-O-O-$ bond is within the limits of 125–210 *kJ*. The energy of quantum absorbed by the investigated peroxides is near 670 *kJ*. Starting from this fact, it can be considered that in result of the compounds photolysis the «hot radicals» are formed. In Table 71 there are values of the excess energy obtained by the radicals under the peroxides photodecomposition. This energy can be directed on the alkoxyradicals reactivity increasing; the spatial interactions are facilitated [158] and the probabilities of their reactions decomposition on smaller radicals and molecules are increased at the expense of the radicals configuration change.

It can be seen from the Table 71 that a part of the secondary radicals forming under the RO decomposition via peroxides photolysis reactions in solid phase is some less.

At the photodecomposition of peroxides $(R_1)_2NCH_2OOC(CH_3)_3$ and $R_1\overset{OR_2}{\underset{H}{C}}-OOC(CH_3)_3$ the primary activated radicals $(R_1)_2NCH_2O$ (I), $R_1\overset{OR_2}{\underset{H}{C}}-O\bullet$ (II) and $OC(CH_3)_3$ (III) are formed. Correspondingly, the radicals (I) and (II) are inactive in reactions of the hydrogen atom break and they can easy decompose on more simple radicals and molecular products.

Table 71. Comparative data of thermal– and photodecomposition of peroxides

Peroxide	Energy of O–O break, kJ/mole	Energy of O–O absorpt., kJ/mole	Excess of energy Q, kJ/mole	a part of radicals or products per mole of the decomposed peroxide							
				photolysis				thermolysis			
				CH_3	R_1	$(CH_3)_2NC\overset{\bullet}{_O}$	$(CH_3)_3COH$	CH_4	R_1H	$(CH_3)_2NC\overset{H}{_O}$	$(CH_3)_3C\overset{H}{_O}$
PTB	159	670	511	0,22	—	—	1,06	0,68	—	—	0,80
I (peroxide)	134	670	536	0,10	—	1,0	0,90	0,22	—	0,91	0,66
III (peroxide)	134	670	536	0,12	—	1,0	0,84	0,10	—	0,63	—
I (peracetal)	167	670	503	0,14	0,37	—	0,83	0,55	0,19	—	0,40
III (peracetal)	167	670	503	0,18	0,26	—	0,70	0,55	0,21	—	0,42
VI (peracetal)	167	670	503	0,39	0,34	—	0,51	0,19	0,19	—	0,31

However, in solid phase more probability of these activated radicals disproportionation in the cell of matrix exists. Undoubtedly, a part of the disproportionation reaction

$$[(R_1)_2NCH_2O] + [OC(CH_3)_3] \rightarrow [(R_1)_2NCHO] + HOC(CH_3)_3 \qquad (96)$$

should be enough high.

The reaction

$$[(R_1)_2NCHO]^* \rightarrow (R_1)_2N\overset{..}{C}=O \qquad (97)$$

can be an additional source of the $(R_1)_2NCO$ radicals.

From the other hand, the fact of radical pairs of alkoxyl radicals' absence during the illumination process of corresponding peroxides says about the definite role of the above considered reactions.

In molecules of peroxides consisting only from the σ–bonds the activation energy is so great that they adsorb only in a field of the vacuum ultraviolet illumination. The quantity of the absorbed energy is so great (~ 670 $kJ/mole$) that the forming photo–activated molecules able to react practically in any direction.

The $-O-O-$ bond is weak and respectively the magnitude of the resonance integral β is low. If to consider the molecules of peroxides under the aspect of the set consisting from the localized two–centred bonds and after that to analyze the interaction between themselves, then the connecting molecular orbital corresponding to the $-O-O-$ bond will be underlying below the molecular orbital of the unshared p–pair which plays a role of the *HOMO*. Antibonding orbital of the $-O-O-$ bond will be playing a role of the *LFMO*. So, for peroxide compounds the $n-\sigma$ electronic transitions should be typical, *i. e.* the electron transition on *LFMO* is strongly antibonding in this place.

From the other hand, the references data and also our own ones say about fact that the investigated peroxide compounds are decomposed into singlet state. This suggests an idea: are these compounds decomposing under the action of *UV* illumination on high oscillating levels of the state S_0, *i. e.* into the «hot» basic state S_0^{*}? If to consider that the internal and intercombinative conversion proceed isoenergetically, then any activated (at the light absorption) molecule which doesn't illuminate the light in the absence of the photoreaction or energy transition hits into the «hot» basic state. Into solutions and solid matrixes the oscillating activation is losing so quickly that the state S_0^{*} can participate only in the quickest from the all possible reactions. It can be expecting that some monomolecular reactions get on well via life time S_0^{*}.

What is the criterion permitting to prove the participation in the reaction of the «hot» basic state? If the photochemical reaction and well–known thermal reactions proceed in the same ways, then probably photo–process proceeds via the high oscillating levels of state. The evidence of fact that the photoreaction proceeds via S_0^{*} will be stringent, if to display that the corresponding thermal reaction is the single process having an activation energy lower than the energy of photo–activation [159].

In our case, when a quantum of the adsorbed energy is enough high (~ 670 $kJ/mole$) and the activation energy of the peroxides decomposition is 126–167 $kJ/mole$, *i. e.* is not great, then, probably, under these conditions the above–

mentioned criterion can be quite realized; *i. e.* the molecules of peroxides can decompose on free radicals into the «hot» state.

Since the difference between the energy of state $S_0{}^*$ and the energy of thermal decomposition is enough great $(670-(126-167) = (544-503)$ kJ, then the probability of the reaction in $S_0{}^*$ should be great and the loss even a great portion of the oscillating energy makes the molecule more reactive.

If the full energy of the molecules is great, and has a lot of internal freedom degrees upon which this energy can be distributed, then the time of the molecule life increases, and the collisions needed for its stabilization can be enough infrequent. On the contrary, if these conditions are not performed the rate of the chemical reaction can be limited by the deactivation processes under collisions. In such cases it is said that the process is limited by the rate of the energy transition.

The rate constants of the peroxides $(CH_3)_2NCH_2OOC(CH_3)_3$ and

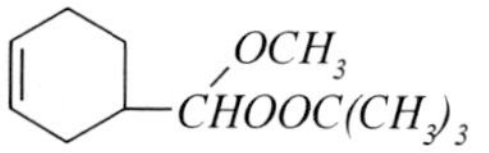

$\bigcirc\!\!-\!\!\overset{\displaystyle OCH_3}{\underset{\displaystyle}{CHOOC(CH_3)_3}}$ photodecomposition can be estimated based on the

following equation [78]:

$$k = A\left(\frac{E - E^{\neq}}{E}\right)^{S-1} \tag{98}$$

where A is the pre–exponential factor of the monomolecular decomposition reaction; E is the energy of the absorbed quantum; $E^{\neq}$ is the activation energy; S is the number of the effective freedom degrees. Let us assume, that $S = 2/3$ of the all freedom degrees into molecule. For the nitrogen–containing peroxide

$$k = 10^{14,5}\left(\frac{669-134}{669}\right)^{47} = 10^{9,3}\,s^{-1} \tag{99}$$

For the peracetal

$$k = 10^{18}\left(\frac{669-159}{669}\right)^{70} = 10^{9,6}\,s^{-1} \tag{100}$$

The time of life is equal to $10^{-9,3}$ s and $10^{-9,6}$ s respectively.

These data are qualitatively agreed with the experimental estimated values of the time of life $10^{-9,3}$ s and $10^{-9,6}$ s respectively. Satisfactory agreement says about fact that for this period of time the peroxide monomolecular decomposition reaction has time into S_0^* state, $i.\ e.$ into the «hot» basic state.

Chemical mechanisms of thermal decomposition into liquid phase and of the photodecomposition into solid phase for the investigated compounds have a lot of common peculiarities. The difference is observed only into the quantitative yield of the final products of decomposition caused by the specificity of the solid phase and by a high reactivity of the primary alkoxyl radicals which can be able by «hot» as a result of their formation from the «hot» molecules of peroxides.

It is seems to us, that some structure specificity of the reactive center for some peroxy compounds during the photodecomposition process should be reflected on the formation of primary products of the decomposition.

Let us consider the nitrogen–containing peroxide $(CH_3)_2NCH_2OOC(CH_3)_3$ and peroxyacetal $\begin{array}{c}\\-\overset{OCH_3}{\underset{CHOOC(CH_3)_3}{\diagup}}\end{array}$. It has been shown early that the bond

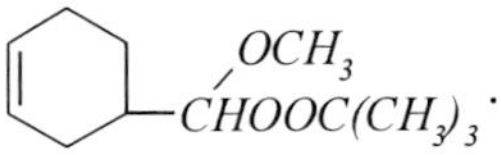

$N...O_\beta$ into $\begin{array}{c} N\cdots\cdots O_\beta{-}C \\ H_2C{-\!-\!-}O_\alpha \end{array}$ (1) is less durable and in any case is less

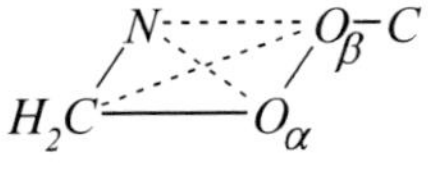

durable than the bond $O_\gamma...O_\beta$ in $\begin{array}{c} R_2O\cdots\cdots O_\beta{-}R \\ C{-\!-\!-}O_\alpha \end{array}$ (2).

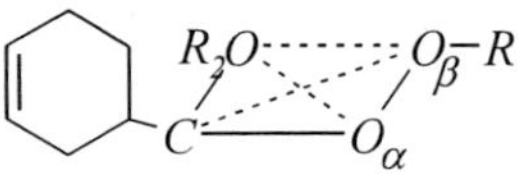

When the systems (1) and (2) absorb in quantum of energy (670 kJ) then the absorbed energy by $-O-O-$ bond is distributed respectively upon the different freedom degree of molecules. It can be assumed that in system (1) as a result of the weak $N...O_\beta$ bond the energy will has the less «freedom» in sense of the energy transition than in a case of (2), where the bond $O_\gamma...O_\beta$ is more durable and is by some additional canal for the energy distribution upon other bonds. In the first case it can be expected the energy concentration on the

fragment $\begin{array}{c} H\diagdown\ \diagup \\ C \\ H\diagup\ \diagdown \end{array}$ of the molecule $\begin{array}{c} H\diagdown\ \diagup N \\ C\cdots\cdots O{-}R \\ H\diagup\ \diagdown O \end{array}$. As a result, one or at once two atoms of hydrogen can be slivered with the formation of free radical $(CH_3)_2NC{=}O$. In molecule (2) the similar concentration of energy is excluded and in connection with this fact it can be realized two ways of the decomposition with the breaking of bonds $O-CH_3$ and $\begin{array}{c}\\+C{-}\end{array}$ into

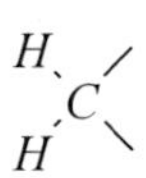

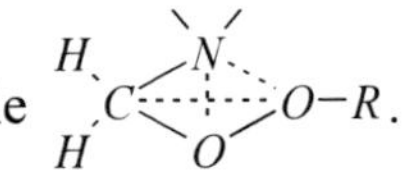

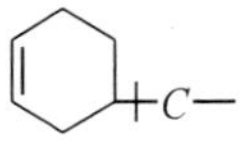

molecule.

Into the activated complex the bonds $N...O_\beta$ and $O_\gamma...O_\beta$, $C...O_\beta$ and $O_\beta-O_\alpha$ are weakened; but at the same time the bonds $N...O_\alpha$ and $O_\gamma...O_\alpha$ are reinforced. This means that the probability of the free «hot» radicals 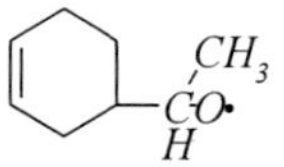 and $(CH_3)_2N-\overset{H}{\underset{H}{C}}-O\bullet$ formation is higher.

Concentration of energy on indicated bonds leads to the formation of corresponding radicals and to the activated $(CH_3)_2NCO$ or $(CH_3)_2NC\overset{O}{\underset{H}{\diagup}}$ which reacting with the $OC(CH_3)_3$ gives the radicals $(CH_3)_2NCO$. The break of the two protons at once in radical $(CH_3)_2N-\overset{H}{\underset{H}{C}}-O\bullet$ can be quite assumed since if to take off the value of the activation energy of nitrogen–containing peroxide decomposition 134 kJ from the 670 kJ then near 272 kJ of energy will be distributed on every from radicals $(CH_3)_2NCH_2O$ and $OC(CH_3)_3$. This energy is enough in order to overcome the potential barrier of the hydrogen atoms break into nitrogen–containing radical. However, it is necessary to take into account the factor of the energy lose by radicals in solid phase and also the disproportionation reaction of radicals which is typical for this phase. Since a lot of *tert*–butyl alcohol is formed under photolysis of $(CH_3)_2NCH_2OOC(CH_3)_3$ so a part of the disproportionation reaction $(CH_3)_2NCH_2O + OC(CH_3)_3$ should be great. The same argumentations are typical also at the consideration of radical 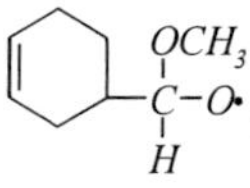 , which can be decomposed also upon two different bonds.

Thus, at low temperatures under the investigated peroxides photolysis the activated alkoxyradicals are formed, reactivity of which in disproportionation reactions, reactions of hydrogen atom breaking and decomposition reaction is higher.

Photodecomposition of peroxides in solid phase proceeds in a state S_0^*.

The value of the effective constants rate for the fixed radicals accumulation at the photodecomposition insufficiently depends on the nature of the substitutes in peroxides of the investigated series. For example, for the compound by $R_1-\overset{OR_2}{\underset{H}{C}}-OOC(CH_3)_3$ structure the accumulation rate for the fixed radicals R_1 is dictated by the R_1-C bond strength and also by the reactions of their decay. Probably, the nature of the substitute R_1 essentially is not reflected on the R_1–C bond strength and the pre–exponential factor is approximately the same and insufficiently depends on the nature of R_1.

The community of the chemical mechanisms of thermolysis in liquid phase and the photolysis in solid phase for the investigated peroxides consists in fact that both of these processes are radical; and the difference between them consists in the nature of the forming free radicals which give in both cases the same final products of reactions, but in different proportions as a result of the different transformation reactions. It is necessary to note, that a great part of these chemical mechanisms appears at the expense of the intra–cell disproportionation reactions of radicals.

ABOUT REACTIVITY OF THE PEROXIDES DECOMPOSITION UNDER THE ACTION OF SOME REACTANTS

In previous Chapters it was considered mainly the reactions of homolytical decomposition of organic peroxides under the action of heat and *UV*–illumination. However, it was interesting to consider some aspects of the peroxide decomposition reactivity under the action of the nucleophilic agents.

7.1. GENERAL INFORMATION ABOUT REACTIONS OF PEROXIDES WITH AMINES

It is well–known, that in a case of the reactions of peroxy compounds with the amines the peroxides bulge as the oxidizing agents and amine as the reducing agents.

The most in detail it has been studied the chemical mechanism of the benzoil peroxide *(BP)* interaction with a series of amines [160–162]. In these works it is noted that the benzoil peroxide can be decomposed both in accordance with the molecular and radical mechanisms.

Horner and *Schwenk* [160–162] hold the opinion that the *PB* interacts with the amines in accordance with the radical chemical mechanism.

Walling [160] supposes both of monomolecular and radical chemical mechanisms simultaneously. Authors of works [163, 164] give preference to the radical mechanism.

It was unambiguously determined the radical chemical mechanism of the *BP* interaction with the triethylamine by *Beyleryan* [165].

Buchachenko and *Pobiedimsky* [166] discovered the cation–radicals of the aromatic amines by *EPR* method. Their scheme includes the stage of the electron transfer from the amine on the peroxide and formation of the ion–radical pair. In acidic and polaric media the cation–radical stabilization proceeds. In non–polar solvents the cation–radical is deproteinizated into the radical products.

An existence of the correlation dependencies of the rates constants for the reactions of amines with the peroxides on the ionization potentials (amines) and on the electron affinity (peroxide) says about the community of chemical mechanism of the peroxides and amines interaction [167–170].

In accordance with the *Mulliken's* theory [171] the transfer of the donor–acceptor pair proceeds between the higher completed molecular orbital of donor and low vacant molecular orbital of acceptor. It is clear, that such interaction will be effective when the energies of the respective orbitals will be enough near upon the values.

Summarizing the above–said, it can be concluded that in references three approaches for the interpretation of chemical mechanisms of peroxides with amines interaction are observed. First is purely radical chemical mechanism [160, 172, 173]; the second is ionic chemical mechanism [174, 175] and the third is mixed from the two above–listed [176, 177]. Which chemical mechanism predominates in the third variant? It depends on the ionization potential of amine, electron affinity of peroxide and the solvent.

In accordance with the *ref.* [174–177] it is necessary to expect, that the low–basic amines should be oxidized mainly accordingly to radical chemical mechanism and the strong–basic amines (namely, secondary aliphatic amines) should be oxidized accordingly to molecular chemical mechanism. Such variety of the assumptions about chemical mechanism of the reactions is explained by different imaginations of authors about the structure of amine–peroxide complex. It is necessary to note that the many reactions are considered by ionic or molecular ones only due to fact that under their proceeding the free radicals are not discovered. In reality, it is possible the variant of the crypto–radical reaction (an interaction in cell with a great rates) [178].

So, in *references* there are enough imaginations about the role of amines by different structure via the interaction with peroxides process. However, the role of the peroxides by different structure is represented under this aspect insignificantly. *The last is the subject of our investigation.*

7.2. STUDIES OF THE REACTION OF PEROXIDES WITH 1,1–DIPHENYL–2–PIKRYLHYDRAZYL (DPhPH)

The more in detail it was investigated the reactions of peroxides with the triphenylverdazyl and phenoxyl radicals [179–182].

An investigation of the reactions of peroxides with the stable radical *DPhPH* it is interesting from the point of view the chemical mechanism of the peroxides–nucleophiles reaction studying.

The following peroxy compounds have been used as the investigation objects:

$(CH_3)_2NCH_2OOC(CH_3)_3$ (*I*);

$C_5H_{10}NCH_2OOC(CH_3)_3$ (*II*);

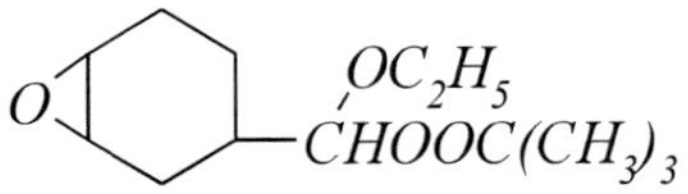

(*III*);

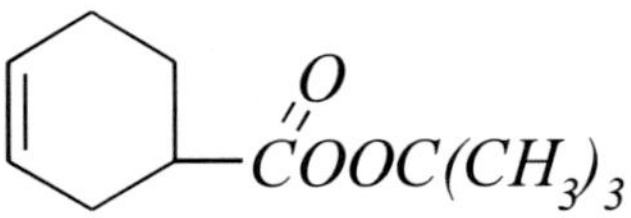

(*IV*);

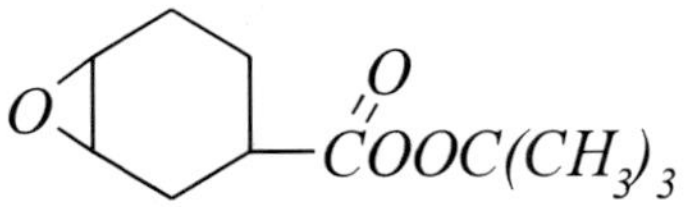

(*V*)

Kinetics of the peroxides with *DPhPH* interaction [183] has been studied at the 100–multiple excess of peroxides. Data on the kinetics of interaction are represented in Table 72. On Figure 57 there are typical kinetic curves of the *DPhPH* consumption in reactions with peroxides. The reaction order per *DPhPH* is the first.

It can be seen from the Table 72, that the reactions proceed with approximately the same activation energy.

For nitrogen–containing peroxides in clorobenzene the rate constants are some higher than for other peroxides. In chloroform this difference in values of the rates constants is practically graded; this is probably connected with the specific influence of the solvent on the reaction rate. For the reactions of peroxyacetals and peresters the lower rates constants are typical. The all of these facts will be explained below. Since the activation energies for the reactions of peroxide with *DPhPH* are approximately the same and the rates constants are respectively the same, then the activation entropy is the responsible factor for the reactions rate.

In chloroform for peroxide *(I)* the activation energy is some increased for quite clear reasons (the solvation of peroxy group and active group of *DPhPH*).

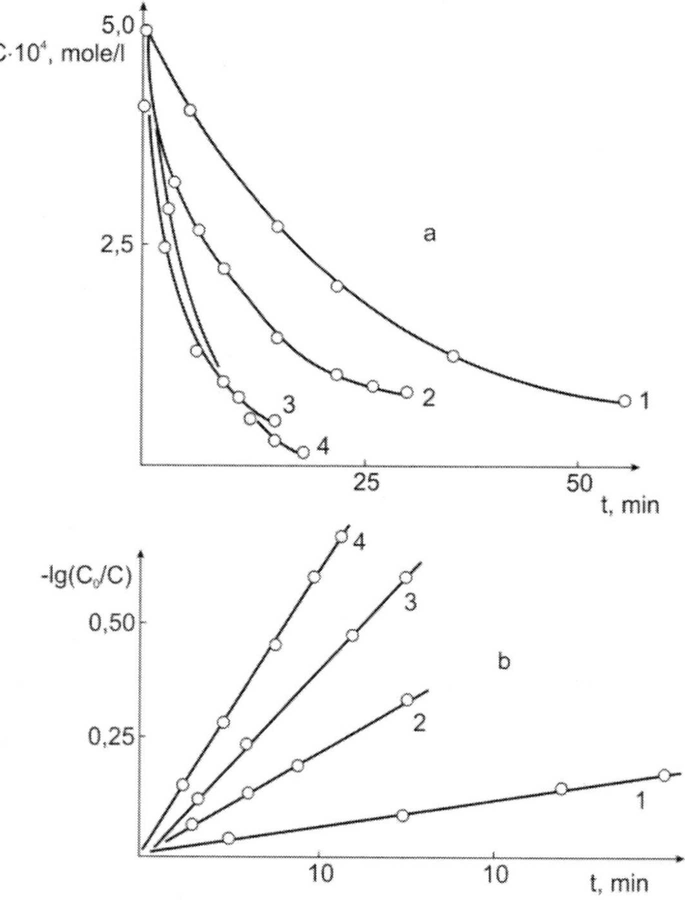

Figure 57. The kinetic curves of the *DPhPH* consumption (*a*) via reaction with the peroxide *(I)* and their semi–logarithmic anamorphosises (*b*) at temperatures 320 *K – 1*; 330 *K – 2*; 340 *K – 3*; 350 *K – 4*. The solvent is chlorobenzene. Starting concentration of peroxide is $5 \cdot 10^{-2}$ *mole/l.*

Table 72. Kinetic parameters of *DPhPH* with peroxides reaction in different solvents in a range of temperatures 320 – 340 *K*

Peroxide	Chlorobenzene				DMSO				Chloroform			
	$\Delta k\pm0{,}2$ $k\cdot10^4$, s^{-1}	$\Delta E\pm4{,}2$ E, $kJ/mole$	A, s^{-1}	$\Delta S^{\neq}$, e.u. $T=330\ K$	$\Delta k\pm0{,}2$ $k\cdot10^4$, s^{-1}	$\Delta E\pm4{,}2$ E, $kJ/mole$	A, s^{-1}	$\Delta S^{\neq}$, e.u. $T=330\ K$	$\Delta k\pm0{,}2$ $k\cdot10^4$, s^{-1}	$\Delta E\pm4{,}2$ E, $kJ/mole$	A, s^{-1}	$\Delta S^{\neq}$, e.u. $T=330\ K$
I	15,4	61,5	$3{,}7\cdot10^5$	−117,2	0,8	62,8	$7{,}6\cdot10^5$	−142,2	16,8	76,6	$8{,}6\cdot10^9$	−75,3
II	11,4	59,8	$1{,}9\cdot10^5$	−129,7	0,9	69,9	$1{,}1\cdot10^7$	−96,2	14,3	67,4	$6{,}8\cdot10^7$	−104,6
III	0,9	55,2	$2{,}8\cdot10^4$	−163,2	0,2	58,6	$2{,}9\cdot10^4$	−163,2	8,4	61,9	$5{,}4\cdot10^6$	−125,5
IV	0,2	60,7	$8{,}4\cdot10^4$	−159,0	0,03	61,5	$1{,}5\cdot10^{-4}$	−173,6	6,7	72,8	$2{,}9\cdot10^8$	−92,0
V	0,2	58,6	$4{,}5\cdot10^4$	−163,2	0,05	62,8	$1{,}8\cdot10^4$	−163,2	11,7	71,1	$2{,}3\cdot10^8$	−92,2

It can be seen from data of Table 72, that the values of the rates constants in some manner depend on the polarity of solvents.

It was proved in *ref.* [184, 185] that the complex formation foregoes to the reaction of benzoil peroxide with triethylamine. We was attempted to discover this complex of peroxide with the *DPhPH* by *NMR*–method. Ditert–butylperoxide *(PTB)* was the suitable object for the investigation of complex formation of the peroxide– *DPhPH* reactions. In this peroxide the protons of the methyl groups are equivalent and the *NMR*–spectrum consists from the one line with the shift of 0,94 *m. p.* relatively to the hexamethyldisiloxane (Figure 58). An addition into solution of the *DPhPH* leads to the essential shift of the *NMR* lines from the protons of the *tert*–butyl groups (1,15 *m. p.*).

Besides, the lines corresponding to the protons of the methyl groups of acetone (2,05 *m. p.*) is displayed into the spectrum; this confirms the formation of *tert*–butoxyl radical.

In order to make sure in fact that the shift of the *NMR*–lines of peroxide (*I*) and *PTB* is caused by the interaction of −*O*–*O*− and *N*–*N* bonds it was done the estimation of the 1,1–diphenylpikrylhydrazine influence on the shift of NMR for *tert*–butyl peroxide. In this case (*DPhPH–H* is not paramagnetic substance) it is observed exactly the same shift (1,15 *m. p.*) as same as in a case of *DPhPH*. This is enough strong evidence of the peroxide–*DPhPH* complex formation.

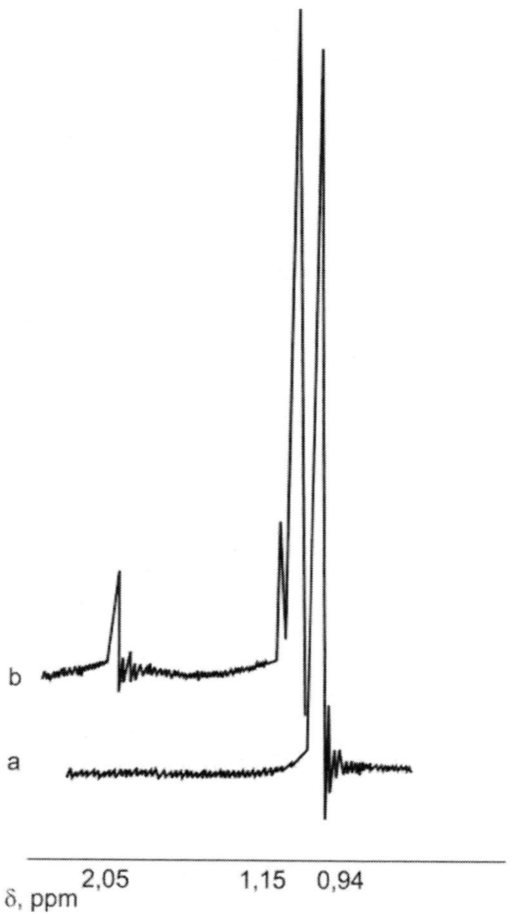

Figure 58. NMR spectrum of *PTB* (*a*); *PTB* + *DPhPH–H* (*b*) in *CCl₄* at 293 *K*. Hexamethyldisiloxane was taken as the standard.

At the interaction of *tert*–butylperbenzoate with *DPhPH* (1 : 1) in chlorobenzene at 313 *K* for 6 hours it was eliminated the not paramagnetic brown product with the melting temperature 440 *K*. The elementary analysis showed that the presented compound has the following structure

$$Ph_2N-N\begin{pmatrix} OC(O)-Ph \\ \\ \end{pmatrix}(NO_2)_3.$$

Besides of this product it was discovered the acetone and *tert*–butanol in quantity practically equal to the balance per peroxide.

At the reaction of *tert*–butylperoxide and *tert*–butylperoxymethyldimethylamine with *DPhPH* in chlorobenzene at 333 *K* it was discovered the compound by following structure $Ph_2N-N\overset{CH_3}{\underset{}{|}}\langle C_6H_2(NO_2)_3 \rangle$.

Summarizing the above–said let us note that the peresters, benzoil peroxide and dialkylperoxides differently interact with *DPhPH*. Forming intermediate complex peroxide–*DPhPH* is decomposed on different final products dependently on the nature of peroxide. In a case of the reaction of benzoil peroxide it can be written the following scheme of reaction:

$$C_6H_5C(O)OOC(O)C_6H_5 + DPhPH \underset{k_2}{\overset{k_1}{\Leftrightarrow}} [X] \qquad (101)$$

$$[X] \xrightarrow{k_3} DPhPH - C(O)C_6H_5 \qquad (102)$$

For dialkyl peroxides the most probable is the following scheme:

$$R_1OOR_2 + DPhPH \underset{k_2}{\overset{k_1}{\Leftrightarrow}} [X] \qquad (103)$$

$$[X] \xrightarrow{k_3} R_1O + R_2O + DPhPH \qquad (104)$$

$$\cdot OR_2 \xrightarrow{k_4} R^{\cdot} + \ inactive \ \ product \qquad (105)$$

$$DPhPH + R^{\cdot} \xrightarrow{k_5} DPhPH - R \qquad (106)$$

$$OR_1 \xrightarrow{k_6} R^{\cdot}{}' + \ inactive \ \ product \qquad (107)$$

$$DhPH + R' \rightarrow DhPH - R \qquad (108)$$

In accordance with the scheme

$$\frac{d[DhPH]}{dt} = k_1[P] - \frac{k_1 \cdot k_2}{k_2 + k_3}[P] \cdot [DhPH] - \frac{k_1 \cdot k_3}{k_2 + k_3}[P] \cdot [DhPH] \qquad (109)$$

Under condition that $P >> DPhPH$ the linear dependence of the $DPhPH$ consumption rate on its concentration should be observed. Really, as we can see from the Figure 59, such dependence is observed.

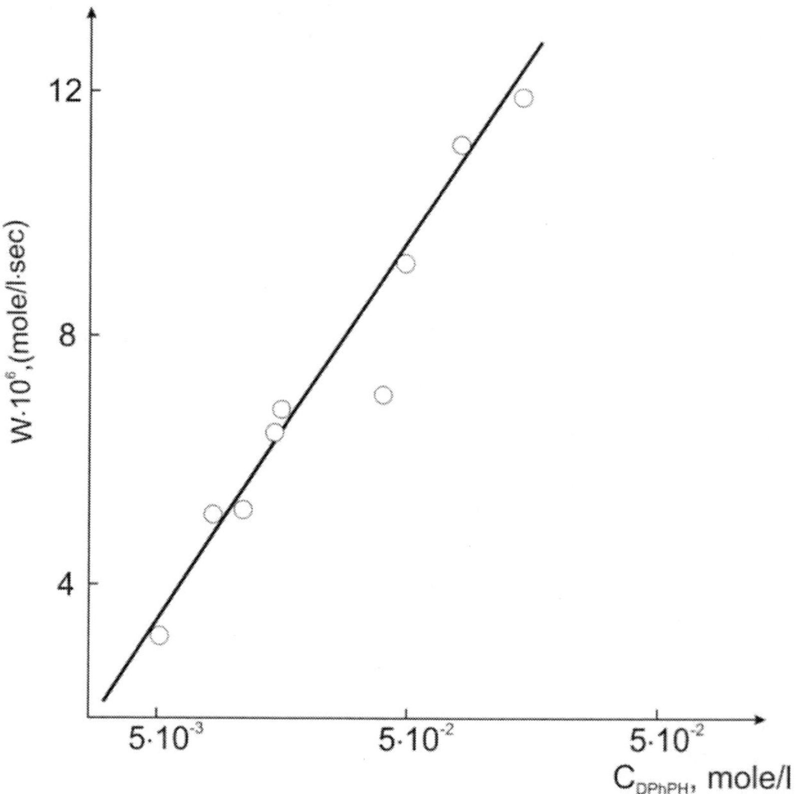

Figure 59. Dependence of the reaction rate for $(CH_3)_2NCH_2OOC(CH_3)_3$ peroxide with $DPhPH$ on concentration of $DPhPH$ in chlorobenzene at 346 K. Starting concentration of peroxide is equal to $2\cdot10^{-1}$ $mole/l$.

7.3. KINETICS OF DPhPH INTERACTION WITH SILICON–CONTAINING PEROXIDES

In *ref.* [184, 185] we have studied the kinetics of *DPhPH* interaction with the silicon–containing peroxides. As the objects of investigation we have studied the compounds by following structure *RSiOOC(CH₃)₃*, where *R* = *(C₂H₅O)(CH₃)₂ (I); (CH₃)₃ (II); (C₂H₅)₃ (III); (C₆H₅O)(CH₃)₂ (IV)* and hexamethyldisilylperoxide *(CH₃)₃SiOOSi(CH₃)₃ (V)*.

The working concentration of *DPhPH* is equal to 0,01 *mole/l.* The reaction kinetics was studied under conditions of the peroxides excess in 10 times in a range of temperatures 293–353 *K.* The reaction order per *DPhPH* is equal to 1.

In Table 73 there are data of kinetic parameters for reaction of peroxides *I–V* with *DPhPH*. As we can see from the data of Table 73, the reaction proceeds with the sufficient low and approximately the same activation energies with the exception of the compound *IV*. The values of the reaction activation entropies are some differed and are in a range of 177,8–221,3 *e. u.* This difference in entropies probably is caused by the structure of the studied objects.

The reaction of *DPhPH* with the silicon–containing peroxides in polar solvents (dimethylsulfoxide, dimethylformamide and acetone) proceeds extremely slowly. In low–polarity solvents (hexane, chlorobenzene) the reaction proceeds at room temperature. Polar solvents make difficult the transfer process on peroxy bond of the unshared pair of the electron of nitrogen atom of *DPhPH*, since the reagents themselves are blocked by the polar molecules of the solvents.

Silicon–containing peroxides as same as the dialkylperoxides via reaction with *DPhPH* form the intermediate complex. On Figure 60 there is *NMR*–spectrum of peroxide P in the presence of *DPhPH*. Chemical shift of the *NMR*–lines from the protons of *tert*–butyl groups is less than in a case of *tert*-butylperoxide (1,15 *m. p.*) and is equal to 1,04 *m. p.*

The influence of the solvent polarity on kinetics of the reaction of peroxide–*DPhPH* interaction can be considered under the *Kirkwood's* approximation [186]. Under this approximation the dependence of the reaction rate constant on dielectric penetrability of the solvent is represented by the expression

$$lg\,k = lg\,k_0 - \frac{2,303}{kT} \cdot \frac{\varepsilon-1}{2\varepsilon+1}\left(\frac{\mu_1^2}{r_1^3} + \frac{\mu_2^2}{r_2^3} - \frac{\mu_{\neq}^2}{r_{\neq}^3}\right) \qquad (110)$$

where k and k_0 are the reaction rates constants in medium with the dielectric penetrability ε and in gaseous phase ($\varepsilon = 1$) respectively; $\mu_{\neq}$, $r_{\neq}$, μ_1, r_1, μ_1, r_1 are dipole moments and radiuses of the activated complex and of the starting reagents (peroxide and *DPhPH*), K is the *Boltzman's* constant; T is temperature.

On Figure 61 the dependence of $lg\,k$ on $\dfrac{\varepsilon-1}{2\varepsilon+1}$ is represented for the peroxide *III* (hexane–tetrahydrofurane).

Table 73. Kinetic parameters of the *DPhPH* with the silicon–containing peroxides reaction in chlorobenzene

Peroxide	$T,\,K$	$k \pm$ $0,1\cdot10^3,\,s^{-1}$	$E_{act.} \pm 4,2,$ $kJ/mole$	$lg\,A$	$\Delta S^{\neq}$ e. u. $T = 313\,K$
I	303	1,1	34,7	2,9	−197,5
	313	1,3			
	323	1,4			
	333	2,8			
II	313	0,6	31,0	2,0	−215,9
	323	0,7			
	333	1,2			
	343	2,5			
III	303	1,1	36,4	3,2	−191,6
	313	1,4			
	333	1,6			
IV	313	1,1	17,6	−0,02	−253,6
	323	1,2			
	343	1,8			
	353	2,1			
V	313	0,1	34,7	1,8	−218,8
	323	0,2			
	393	0,3			
	343	0,4			

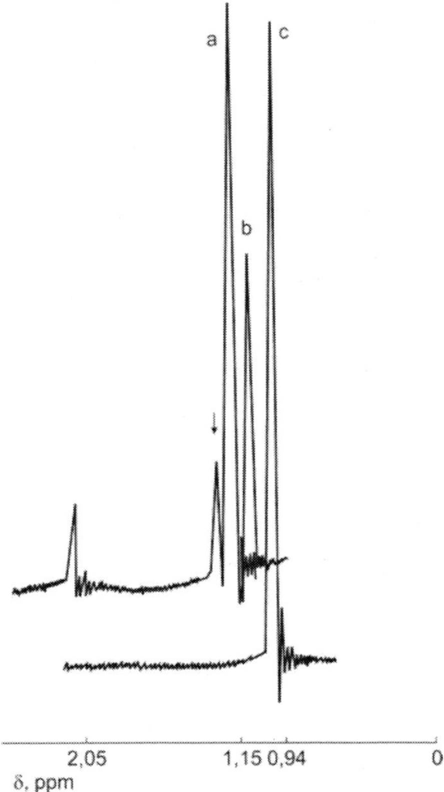

Figure 60. *NMR–spectrum of PTB (a); peroxide P + DPhPH (b); PTB + DPhPH (c) in CCl₄ at 293 K.*

Knowing the values of r and μ for the starting reagents and $r_{\neq}$ for the activated complex it can be estimated the dipole moment of this complex. The values of r_1 and r_2 are determined by the expression $r^3 = \dfrac{3V}{4\pi N_A}$, where V is the molecular volume, N_A is the *Avogadro's* number; $r_1 = 4{,}6\ \overset{0}{A}$; $r_2 = 5{,}6\ \overset{0}{A}$; $r_{\neq}^3 = r_1^3 + r_2^3$ [185]; $\mu_1 = 1{,}8\ D$ (in benzene at 298 K) and $\mu_2 = 4{,}92\ D$ [187]. Estimated value of the dipole moment $\mu_{\neq}$ of activated complex is equal to 9,4 D; *i. e.* this complex is sufficiently strongly polar formation. A great value of $\mu_{\neq}$ is the forcible argument in favour of proceeding the studied reaction with the charge transfer.

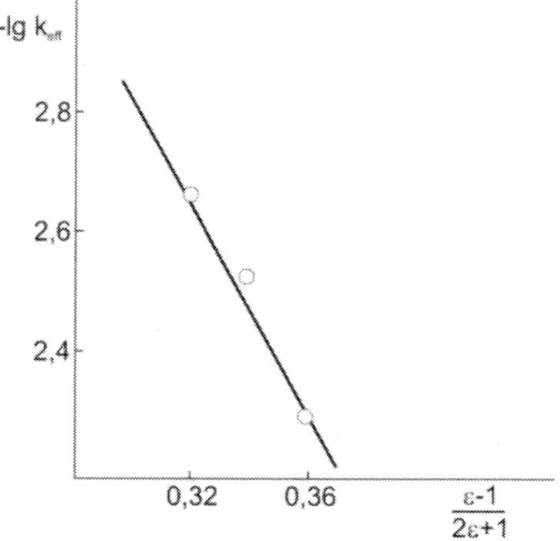

Figure 61. Dependence of the logarithm effective rate constant for the reaction of *DPhPH* with peroxide *III* on the polarity of medium.

As to the chemical mechanism of the silicon–containing peroxides with *DPhPH* interaction it can be done the conclusion about its radical nature based on the data of Table 74. The quantitative formation of the

product is caused by the reaction of radicals *DPhPH* and CH_3 in cell of the solvent. The formation of other products *(see* Table 74*)* can be represented only via the radical reactions way.

The reaction scheme of the silicon–containing peroxides with *DPhPH* interaction is as follows:

$$(R_1)_3 SiOOR + DhPH \underset{k_2}{\overset{k_1}{\Leftrightarrow}} [X] \qquad (111)$$

$$[X] \xrightarrow{k_3} OR_2 + (R_1)_3 SiO + DPhPH \qquad (112)$$

$$OR_2 \xrightarrow{k_4} R_2' + \quad inactive \quad product \qquad (113)$$

Table 74. Data of the chromatographic analysis *(mole/l)* for reaction of peroxide III with *DPhPH* in some solvents under different starting concentrations of the reagents *($C_{peroxide}$, C_{DPhPH})* *(T = 333 K)*

№	$C_{peroxide}$	C_{DPhPH}	Solvent	$(CH_3)_2$ CO	$(CH_3)_3$ COH	$(C_2H_5)_3$ $SiOSi(C_2H_5)_3$	$(C_2H_5)_3$ $SiOH$
1	0,07	0,01	DMFA	0,01	—	—	—
2	0,10	0,05	DMFA	0,04	0,01	0,01	0,01
3	0,10	0,10	DMFA	0,10	0,02	0,03	0,03
4	0,23	0,19	DMFA	0,17	0,04	0,08	0,06
5	0,10	0,01	ChB	0,01	—	—	0,01
6	0,20	0,04	DMFA	0,01	0,03	0,01	0,02

$$DPhPH + R_2' \xrightarrow{\ k_5\ } DPhPH - R_2' \tag{114}$$

$$OR_2 + RH \xrightarrow{\ k_6\ } HOR_2 + R^{\cdot} \tag{115}$$

$$(R_1)_3 SiO + RH \xrightarrow{\ k_7\ } (R_1)_3 SiOH + R^{\cdot} \tag{116}$$

$$2(R_1)_3 SiOH \xrightarrow{\ k_8\ } (R_1)_3 SiOSi(R_1)_3 \tag{117}$$

7.4. INVESTIGATION OF REACTION OF 1,1–DIPHENYL–2–PIKRYLHYDRAZIN (DPhPH–H) WITH PEROXIDES

It has been studied the reaction of *DPhPH* with peroxides

$$(CH_3)_2NCH_2OOC(CH_3)_3 \quad (I), \qquad n.\text{-}C_4H_9\overset{O}{\overset{\cdot\cdot}{C}}OOC(CH_3)_3 \quad (II),$$

$O\langle\rangle\text{-}\overset{O}{\overset{\|}{C}}OOC(CH_3)_3$ *(III)*, $C_6H_5\text{-}\overset{O}{\overset{\|}{C}}OOC(CH_3)_3$ *(IV)* under conditions of the *DPhPH–H* concentration in 10–multiple excess. The concentration of peroxides were within the limits of 10^{-2} *mole/l*.

On Figure 62 there are kinetic curves of the product accumulation of reaction – *DPhPH*. General order of the reaction upon reagents is equal to 2.

Kinetic parameters of the *DPhPH–H* with peroxides interaction are represented in Table 75. Energies and activation entropies for these reactions are sufficiently strongly depend on the structure of the peroxide compounds *(see* Table 75).

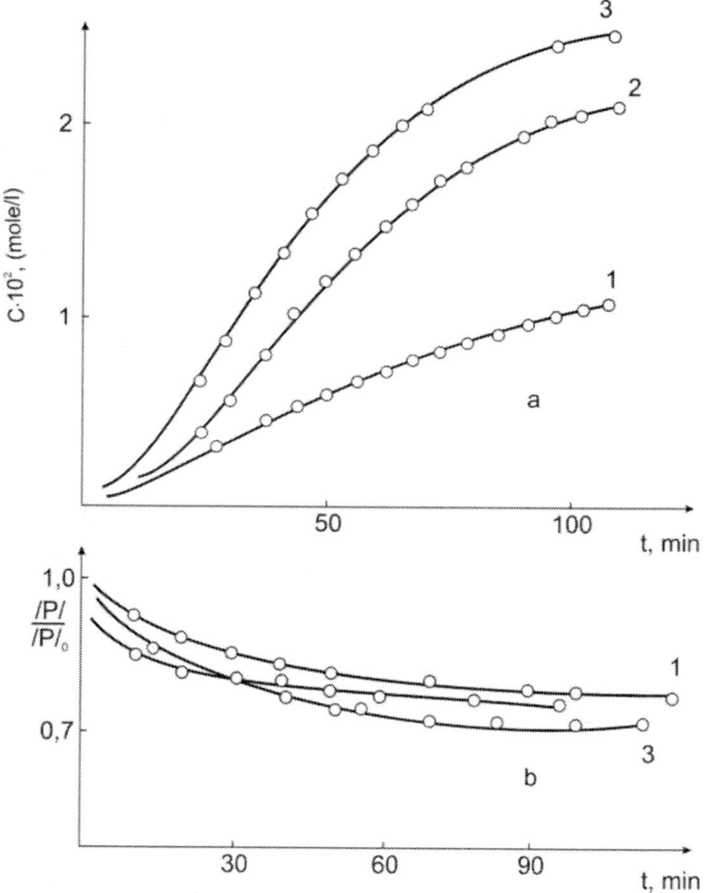

Figure 62. Kinetic curves (*a*) of the radicals accumulation (*DPhPH*) at the interaction of peroxide $C_4H_9\overset{\cdot}{C}\!-\!OOC(CH_3)_3$ with *DPhPH–H* at 313 *K* (1), 323 *K* (2) and 333 *K* (3). The decomposition of peroxides (*I* – 1; *II* – 2; *III* – 3) via reaction of *DPhPH–H* in chloroform (*b*).

Formation of the intermediate complex and the reaction product *tert–*butanol at the reaction of peroxide *I* with *DPhPH–H* in chlorobenzene gives some principles to suppose that the primary act of the reagents interaction can be represented by the following scheme:

$$R-\overset{O}{\overset{\|}{C}}-\underset{(CH_3)_3C-O}{O} + \underset{O_2N \diagdown NO_2 \diagdown NO_2}{\overset{Ph\diagdown N \diagup Ph}{\overset{|}{\underset{|}{N-H}}}} \underset{k_2}{\overset{k_1}{\rightleftarrows}} R-\overset{O}{\overset{\|}{C}}-\underset{(CH_3)_3C-O}{\overset{\ominus}{O}} \quad \underset{O_2N \diagdown NO_2 \diagdown NO_2}{\overset{Ph\diagdown N \diagup Ph}{\overset{\oplus}{\underset{|}{N-H}}}}$$

$$`(118)$$

I doubt whether the presented reaction can be considered only as the reaction of the hydrogen atom transfer from the atom of nitrogen in *DPhPH–H* since this contradicts to fact that the intermediate complex the same as *DPhPH–peroxide* complex is formed with the *DPhPH–H*. It is follows from this that the unshared pair of atom of nitrogen is touched via the reaction process. The other question does the reaction proceed with partial or full transfer of charge?

The formation of *DPhPH* via the reaction process in some manner contradicts to the affirmation about the full transfer of charge.

It is interesting to consider the kinetic data of the reaction of peroxides with the *DPhPH–H* from the point of view the structure of the reactive centre of these peroxy compounds. Let us stop more in detail on the kinetic data of the reaction in chlorobenzene *(see* Table 75*)*. First is fact that the energies and the activation entropies for the reaction of peroxide with *DPhPH–H* are in enough wide diapason. Generally it is observed the following tendency: the stronger are bonds $X...O_\alpha$, $X...O_\beta$, $C...O_\beta$ and $O–O$ in the peroxy compound with the polycyclic center the more is the activation energy of the process for this peroxide and *DPhPH–H*. For the compound $n.-C_4H_9\overset{O}{\overset{\|}{C}}OOC(CH_3)_3$ the energy upon these bonds is the greatest (–1,0288 *a. u.*), for peroxide $(CH_3)_2NCH_2OOC(CH_3)_3$ is smaller (–0,9283 *a. u.*) and for compound $O \diagdown \hspace{-0.5em} \bigcirc \hspace{-0.5em} -\overset{O}{\overset{\|}{C}}OOC(CH_3)_3$ is the least (–0,8952 *a. u.*).

At the consideration of activation entropies for these reactions we can see, that the greater activation entropy of reaction for peroxides (*I*) and (*II*) is caused by the higher «inflexibility» of the starting reagents, *i. e.* we will be called by the strength of the cyclic center of peroxide. Absolutely contrary situation is observed for the kinetic parameters of peroxides (*III*) and (*IV*) with *DPhPH–H* reaction.

Summarizing the all above–said it can be done the following qualitative conclusions as to reactions of the bimolecular interaction of the peroxides with the cyclic reactive center having the nucleophile: firstly, kinetic parameters of the reaction will be caused by the stabilization energy of the peroxides, *i. e.* by the strength of the $X...O_\alpha$, $X...O_\beta$, $C...O_\beta$ and $O–O$ bonds. The stronger these bonds, the greater is energy and activation entropy of the process; secondly, from the first conclusion it is follows the compensatory dependence between the entropy and the activation energy for the bimolecular reactions of peroxides. It is clear, that these conclusions are qualitative and should be checked on the multiple others reactions of peroxides in order to obtain of any quantitative fundamentals.

7.5. INTERACTION OF NITROGEN–CONTAINING PEROXIDES WITH A–NAPHTHYLAMINE

Tert–butylperoxymethyldiethylamine (*I*) and *tert*–butylperoxymethyldiisopro-pylamine (*II*) were used as the investigation objects.

The reaction has been studied at the sufficiently high temperatures (393–403 *K*) in chlorobenzene at the excess of the α–naphthylamine.

On Figure 63 there is dependence of the starting rate of the peroxides (*I*) and (*II*) consumption on their concentration under condition *[InH] >> ROOR*$_1$. It follows from this, that the reaction order per the peroxide is equal to 1. It follows also from the Figure 64 that the reaction has the first order per the inhibitor. Hence, the general reaction order per the reagents is equal to 2.

Table 75. Kinetic parameters of the *DPhPH–H* with the peroxides interaction

Peroxide	Chlorobenzene					Dimethylphtalate				
	T, K	$\Delta k\pm0,1$ $k\cdot10^4, s^{-1}$	$\Delta E\pm4,2$ $E,$ $kJ/mole$	$A,$ s^{-1}	$\Delta S^{\neq}, e.\,u.$	T, K	$\Delta k\pm0,1$ $k\cdot10^4, s^{-1}$	$\Delta E\pm4,2$ $E,$ $kJ/mole$	$A,$ s^{-1}	$\Delta S^{\neq}, e.\,u.$
I	343	0,4	87,9	10^9	−80,3	332	2,0	60,7	$1,3\cdot10^6$	−138,1
	352	0,8				343	6,8			
	363	2,3				353	10,8			
II	343	0,2	107,1	$2,7\cdot10^{11}$	−38,9	343	2,2	47,3	$3,6\cdot10^3$	−182,0
	363	0,7				353	3,2			
	373	1,4				363	5,6			
III	333	0,4	75,6	$2,3\cdot10^7$	−114,2	333	1,7	64,9	$2,8\cdot10^6$	−131,4
	343	0,7				343	3,5			
	353	1,6				353	6,1			
IV	303	2,4	43,9	$5,8\cdot10^3$	−173,6	303	1,8	30,1	$1,7\cdot10^2$	−227,2
	313	4,1				313	2,7			
	323	7,4				323	3,8			
	343	11,3				343	42,3			

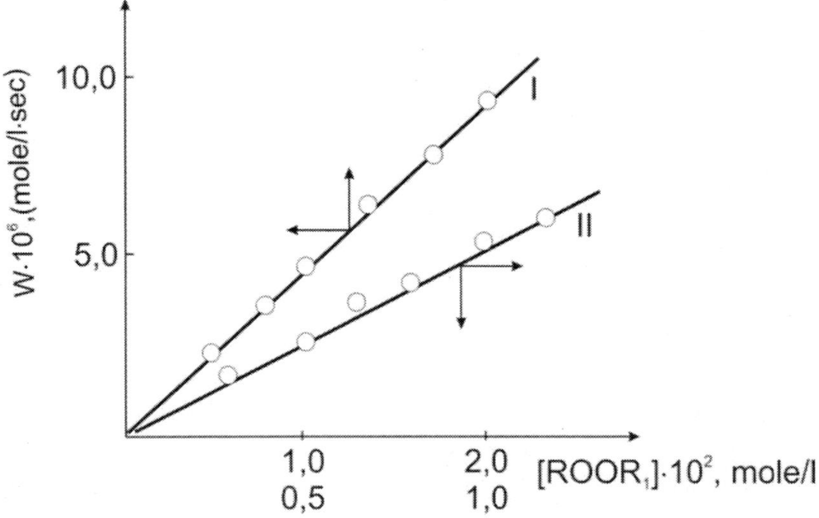

Figure 63. Dependence of the decomposition rate for dimethylaminemethyl–*tert*–butylperoxide (*I*) and diisopropylaminomethyl–*tert*–amylperoxide (*II*) on concentration in the presence of inhibitor in chlorobenzene. *[InH] >>[ROOR₁]·[InH]–const.*

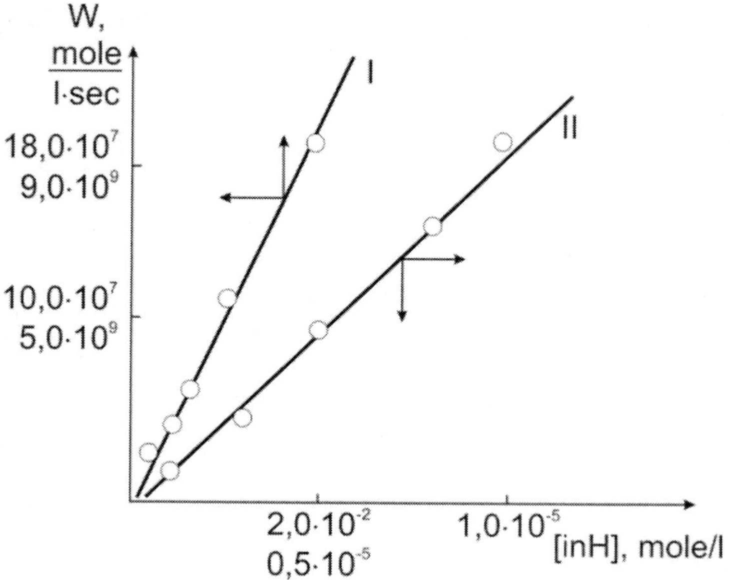

Figure 64. Dependence of the decomposition rate constants for dimethylaminomethyl–*tert*–butylperoxide (*I*) and diisopropylaminomethyl–*tert*–amylperoxide (*II*) on concentration of inhibitor in chlorobenzene under condition *[InH] >> [ROOR₁]* · *[ROOR₁] – const.*

General rate *(W)* of the peroxides (*I*) and (*II*) interaction with the α–naphthylamine at more higher temperatures can be represented by the expression $W = W_1 + W_2$ with taken into account of their thermolysis:

$$W_1 = k_1[R_1OOR], \quad W_2 = k_2[ROOR'][InH] \tag{119}$$

Drawing the dependence into coordinates $\dfrac{W}{[R_1OOR']} - [InH]$ it can be estimated the k_1 and k_2. Such dependence for peroxides (*I*) and (*II*) is represented on Figure 65 and Figure 66. Calculated values for (*I*) – $4,7 \cdot 10^{-3}$ *l·mole⁻* The values of activation energy and pre–exponent for the bimolecular reaction (*I*) within the interval of temperatures 388–408 *K* are equal to 47,7 *kJ/mole* and $8,5 \cdot 10^4$ *l·mole⁻¹·s⁻¹* respectively. Let us note, that the values of the rates constants k_1 for peroxides (*I*) and (*II*) are respectively equal to $2,4 \cdot 10^{-4}$ *s⁻¹* (399 *K*) and $2,6 \cdot 10^{-4}$ *s⁻¹* (403 *K*); this is in good agreement with the same data obtained under the peroxides (*I*) and (*II*) decomposition in styrene $(2,4 \cdot 10^{-4}$ *s⁻¹* and $3,0 \cdot 10^{-4}$ *s⁻¹*).

$^1 \cdot s^{-1}$ (*T* = 399 *K*), (*II*) – $14,5 \cdot 10^{-3}$ *l·mole⁻¹·s⁻¹* (*T* = 403 *K*).

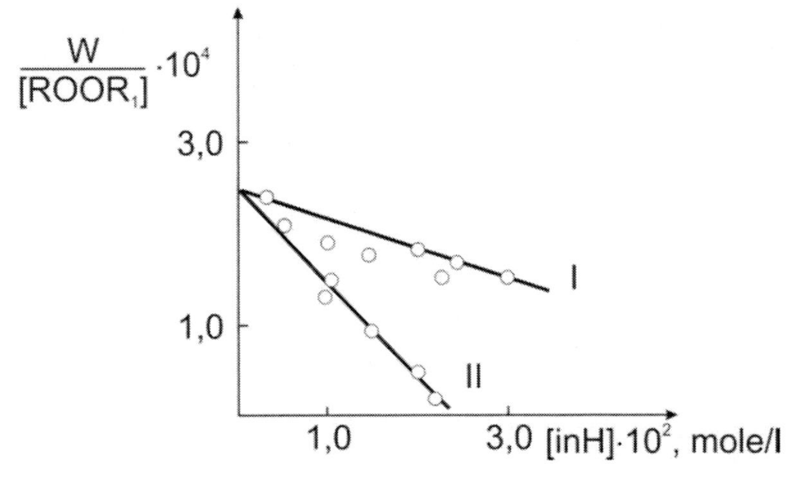

Figure 65. Dependence of $\dfrac{W}{[ROOR_1]}$ on *[InH]* under decomposition of diethylaminomethyl–*tert*–butylperoxide (*I*) (399 *K*) and diisopropylaminomethyl–*tert*–amylperoxide (*II*) (403 *K*) in chlorobenzene at the condition *[InH] ≤ [ROOR₁]*.

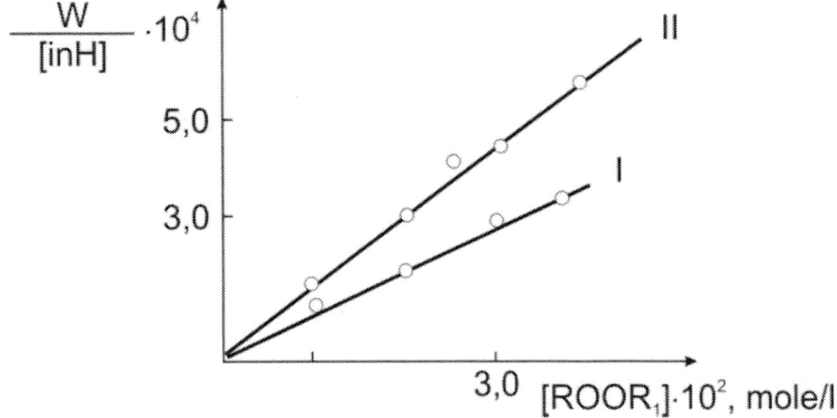

Figure 66. Dependence of $\dfrac{W}{[InH]}$ on *[ROOR₁]* under decomposition of diethylaminomethyl–*tert*–butylperoxide (*I*) (399 *K*) and diisopropylaminomethyl–*tert*–amylperoxide (*II*) (403 *K*) in chlorobenzene.

7.6. ABOUT INTERACTION OF TERT.– BUTYLPERBENZOATE WITH A–NAPHTHYLAMINE

The reaction of *tert*–butylperbenzoate *(TBPB)* with α–naphthylamine has been studied insufficiently and even there are assumptions [188] about the absence of such interaction.

In *ref.* [189] it has been shown by us that in the presence of 80–multiple excess of α–naphthylamine in a range of temperatures 333–353 K in chlorobenzene it is observed a great consumption of peroxide. The reaction rate is near 10^{-6} *mol/l·s*. The general order of reaction is equal to 2. However, even at 10–multiple excess of amine at $T = 383$ K the order is equal to 1,5 (per peroxide); this is an evidence of the peroxide chain decomposition.

In Table 76 there are kinetic parameters of the *TBPB* thermal decomposition in chlorobenzene and *n*–butanol. Concentrations of reagents are $1·10^{-2}$ *mole/l* and $8·10^{-1}$ *mole/l* for peroxide and amine respectively. Probably, the kinetic data of Table 76 are characteristics of the bimolecular interaction of peroxide–amine since the *TBPB* decomposition rate in a range of temperatures 323–353 K is enough low 10^{-9} *mol/l·s* accordingly to [188].

For the temperatures 363–383 K the activation energy of a process is as same as in a case of the lower temperatures. Evidently, the endowment of the thermolysis reaction is insignificant at higher temperatures.

Table 76. Kinetic parameters of the *TBPB* reaction with α–naphthylamine (α–NA) ($C_{TBPB} = 1·10^{-2}$ *mole/l*; $C_{\alpha-NA} = 8·10^{-1}$ *mole/l*)

T, K	Chlorobenzene		Butyl alcohol	
	$k \pm 0{,}24, s^{-1}$	$E \pm 0{,}7,$ *kJ/mole*	$k \pm 0{,}21, s^{-1}$	$E \pm 0{,}9,$ *kJ/mole*
323	—	—	$0{,}54·10^{-5}$	
333	$0{,}44·10^{-4}$		1,46	79,5
343	0,73	49,4	3,12	
353	1,23		—	
363	4,46		$0{,}96·10^{-4}$	99,5
373	6,68	51,8	2,31	
383	10,52		5,38	

However, at less excesses of α–naphthylamine and higher temperatures of reaction in all probability it is necessary to take into account the endowment of the *TBPB* thermolysis reaction.

Some activation energy increasing for interaction of *tert*–butylperbenzoate with α–naphthylamine in *n*–butanol in comparison with the chlorobenzene can be related to the interaction of associate with α–naphthylamine–alcohol and peroxide–alcohol. It is quite clear that the transfer of electron of the unshared pair of nitrogen atom on the vacant orbital of the *O–O* oxygen bond is complicated under presented conditions.

Experimental investigations of the reaction of peroxides with *DPhPH* lead to the assumption that the chemical mechanisms of the interaction for peroxides by different nature in these reactions have the radical character. An interaction of the benzoil peroxide or peresters with *DPhPH* can be considered as the unidirectional nucleophilic reaction, the first stage of which is the formation of the intermediate complex transforming into compound

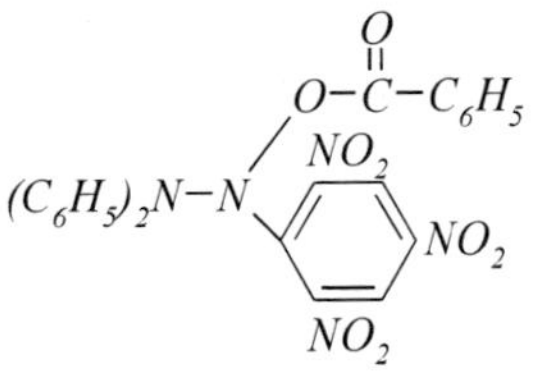

$$(C_6H_5)_2N-N\left\langle \begin{array}{c} O-\overset{\overset{O}{\|}}{C}-C_6H_5 \\ NO_2 \\ \end{array} \right\rangle NO_2 \quad\text{(ring with } NO_2\text{)}$$

as a result of the complex decomposition on free

radicals $C_6H_5\overset{\overset{O}{\|}}{C}-O^{\bullet}$ and *DPhPH* with their following recombination inside of the solvent cell. This chemical mechanism of the *DPhPH* interaction is more visibly discovered in reactions with dialkyl peroxides (di–*tert*–butylperoxide), dialkyl–substituted peroxides $(CH_3)_2NCH_2OOC(CH_3)_3$ and

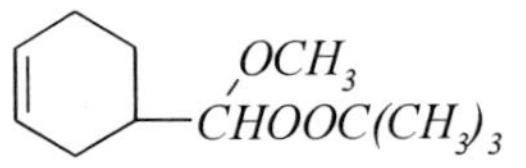

$$\left\langle \begin{array}{c} OCH_3 \\ -CHOOC(CH_3)_3 \end{array} \right.$$

or silicon–containing peroxides. As a result of these compounds reaction with DPhPH the following product is formed

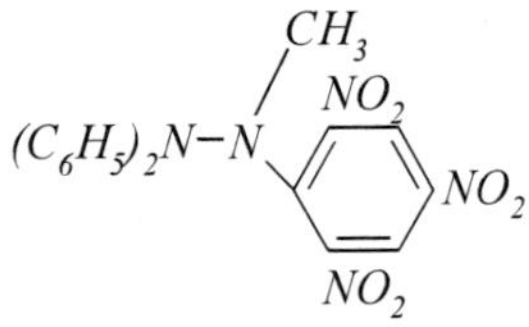

$$(C_6H_5)_2N-N\left\langle \begin{array}{c} CH_3 \\ NO_2 \\ \end{array} \right\rangle NO_2 \quad\text{(ring with } NO_2\text{)}$$

. Its formation can be presented as the result of the intermediate complex decomposition on free alkoxy radicals which detach the radicals CH_3, which can be able to recombine with the *DPhPH*. Probably, forming in the presented cases alkoxyl radicals can not be able recombine with the *DPhPH* as a result of their less electron affinity (–0,15 *a. u.*) than the radical $C_6H_5\overset{\nearrow O}{C}-O\cdot$ (–0,03 *a. u.*).

It seems to us that the primary act of the *DPhPH* reaction with the peroxides is not reduced to the full transfer of the unshared pair of the electron of nitrogen atom on peroxy bridge but only to the partial one with the formation of the unstable intermediate complex which is decomposed on radical products. The itself radical of *DPhPH* plays the role of the activating agent conducive to more easy homolytical destruction of the peroxy bond.

DPhPH–H reacts with the peroxides in accordance with the same chemical mechanism. However, once more additional bond of hydrogen of *DPhPH–H* with one among oxygen of peroxy bridge is formed under the complex formation. During the intermediate complex decomposition this atom of hydrogen overs to the oxygen atom of peroxide with the formation of compound by *ROH* type.

If the process of peroxides reaction with the nitrogen–containing compounds to consider from the point of view the theory of *MO (HMO)* disturbance, then this can be represented as the electrons transfer from the *HOMO* of the nitrogen atom on the *LFMO* of the peroxy bridge. The stabilization energy of the intermediate complex can be estimated accordingly to the usual equation of *HMO*. In a range of the reactions of peroxides with any amine–containing compound this energy should be depend on the value of *LFMO* energy of peroxide since the energy of *MO* of *HOMO* of nucleophile is the constant magnitude. In turn, the stabilization energy should be symbate to the activation energy for this series of reactions. Really, the energy of *LFMO* for silicon–containing compounds is equal to 0,1036–0,1140 *a. u.*, and the activation energy of *DPhPH* reaction with these compounds is equal to *33,5 kJ/mole*, whereas for the peroxides *(CH₃)₂NCH₂OOC(CH₃)₃* and $C_6H_5\overset{\nearrow O}{C}-OOC(CH_3)_3$ the *MO* energy is equal to 0,1720 *a. u.* and 0,1696 *a. u.* and the activation energy is near 58,6 *kJ/mole*.

For the compound O⟨benzene ring⟩$-\overset{\nearrow OCH_3}{C}HOOC(CH_3)_3$ the energy of *MO* is equal to 0,1640 *a. u.* and the activation energy of reaction with *DPhPH* is equal to 55,2 *kJ/mole*.

Among reactions of *DPhPH–H* with peroxides $C_6H_5\overset{O}{\overset{|}{C}}-OO\overset{O}{\overset{|}{C}}-C_6H_5$,

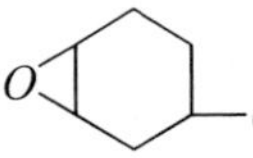

$O=\!\!<\!\!\bigcirc\!\!>\!\!-\overset{O}{\overset{\|}{C}}OOC(CH_3)_3$, $(CH_3)_2NCH_2OOC(CH_3)_3$, $n.-C_4H_9\overset{O}{\overset{|}{C}}OOC(CH_3)_3$ the

MO energies of *LFMO* are equal to 0,1437 *a. u.*, 0,1664 *a. u.*, 0,1696 *a. u.* and 0,1831 *a. u.* The values of activation energies are respectively equal to 41,8 *kJ/mole*, 75,3 *kJ/mole*, 87,9 *kJ/mole* and 107,1 *kJ/mole*.

For the reactions of peroxides $R_1R_2NCH_2OOC(CH_3)_3$ and $C_6H_5\overset{O}{\overset{|}{C}}-OOC(CH_3)_3$ with α–naphthylamine the activation energies are near and equal to 47,7 *kJ/mole* and 50,2 *kJ/mole*. However, the energies of *LFMO* are near also (0,1696 and 0,1720 *a. u.*).

As to the questions of the intermediate complex stabilization and clarification of its nature, this requires the additional experimental investigations carrying out and the calculations of its stabilization energy.

CONCLUDING REMARKS

Developed in the presented work approach to the relationship between the structure of complicated organic peroxides and their reactivity in homolysis reactions with the aim of the kinetic parameters prognostication consists in the interpretation of the experimental data on quantum–chemical language. The essence of this approach is lied in fact that the stabilization energy is proposed to be taken as the index of the reactivity for reactions of the peroxides homolysis; *in other words*, the orbital energy is proposed to be used as the characterizing index not only of the *O–O* bond strength but taking also into account the interaction of reactive centre atoms valence free between themselves.

Covering of any other chemical mechanisms of peroxides decomposition on the homolysis should be discovered in the disparity of the experimentally founded kinetic parameters with the calculated ones. This can be criterion of the complicated chemical mechanism of the peroxides decomposition.

An existence of the peroxides in several isomeric forms is reflected on the values of their kinetic parameters of homolysis. The difference in activation energies of the isomers homolysis consists of not more than 8–9 *kJ/mole*. So, at the prognostication of the kinetic parameters this fact can be taken into account. Sometimes expecting of this, it is necessary to do the conformational analysis of the peroxide molecule.

Evidently, under peroxides homolysis into liquids it is necessary to take into account the volumetric effects of the substitutes. An increase of the substitutes volume leads to the less probability of free oxy–radicals outcome from a cell accordingly to the diffusion chemical mechanism and in such a way facilitates to the decreasing of the activation energy value.

When the stabilization energy of the reactive centre of peroxides atoms is enough great magnitude, then even the solvent able to form the hydrogen bond hasn't the influence on the kinetics of this peroxide. This is explained by the effects of the intramolecular interaction of reactive centre atoms.

Generally, the thermal stability of organic peroxides is considerably caused by the effects of intramolecular interaction of atoms, especially of atoms by $\diagdown C_1\!\!-\!\!O\!\!-\!\!O\!\!-\!\!C_2\diagdown$ centre. Using the experimental and theoretical investigations of the organic peroxides by general $R_1R_2C\overset{X}{-}O\!-\!O\!-\!R$ structure, it was shown, that the most energetically suitable is the screened *(cis)* or *hauche*–conformation of $C\!-\!X$ bond with respect to the $O\!-\!O$ bond. Quantum–chemical interpretation of this phenomenon consists in a benefit of the orbital energy of valence free atoms interaction of the reactive center $(C_1...O_\beta, X...O_\alpha, X...O_\beta)$.

The transition of molecules from the *trans*–conformation in more suitable *cis*–conformation is considered as the intramolecular disturbance. The wave function of the system at the transition into *cis*–form should include the once activated configurations with the charge transfer inside of molecule atoms. Thus, it can be assumed, that the stabilization of *cis*–conformation is determined at the expense of the energy of the charge transfer. Actually, this energy is the stabilization energy of the stable conformation of the peroxide molecule. The energy of electrostatic interactions of center atoms for *trans*–conformer is some less, however this form plays back into the orbital energy.

The stabilization energy, *i. e.* the orbital energy of valence free atoms with the oxygen atom with unshared electron can be the reactivity index of free oxy radicals in different transformation reactions. This means that the reactivity of complicated radicals is determined by their conformation. If the non–valence interactions of the atoms of radical with the oxygen atom are enough great then the reactivity in reactions *(for example,* hydrogen atom break, addition reactions) will be little, and in other reactions (isomerization reactions and *ect.*) will be great.

Considered into presented monograph variant of an estimation of the non–valence interactions of atoms into the peroxide molecule is far from the uncovering of the physical essence of the interaction and that why is presented as the empirical index of the reactivity (stabilization energy). However, even this gives the possibility to understand the reasons of the peroxides conformers' existence, to explain the regularities of the relationship between the peroxides and their reactivity in different reactions, to estimate theoretically the kinetic parameters of peroxides, to explain the low and high values of the pre–exponential factors in reactions, to do the useful conclusions about free or hindered rotation of groups into molecules, to understand the reasons of different stabilization of the activated complex in homolysis reactions, to explain the existence of the compensatory dependence between E and lgA for the reactions of peroxides and, at last, to interpret the reactivity of the oxy radicals in different reactions of their transformation.

Proposed classification of the organic peroxides taking into account the interaction of atoms of the reactive center is conventional, but impartial upon its nature, since is based upon the modern electronic imaginations. It is useful both on the practice and facilitates to the development of the principles of the purposeful synthesis of the peroxides. By introduction of any heteroatom or heteroatom with the different substitute in β–position with respect to the atom of α–oxygen of peroxide bridge it can be widely varied the thermal stability of the peroxides and practically synthesized any peroxides with given estimated kinetic parameters. The analysis of different oxy radicals reactivity into proposed by us variant shows that qualitatively is possible to forecast the radical formation in peroxides homolysis reactions. Finally, this permits in advance to determine also the final products of the peroxides decomposition; this is very important for the practice, where the peroxides are applied as the source of the free radicals.

It is necessary to note that only the central question of the purposeful synthesis of peroxides is solved in presented monograph, *namely* the question of relationship between their electronic structure and reactivity into homolysis reactions for determined type of the objects. However, for the development of the recommendations concerning to the purposeful synthesis of peroxides it is necessary also to solve a series of other questions. First of all this is discovering of the quantitative ratios between the electronic structure and their reactivity and determination of regularities for the relationship between the electron structure of peroxides and their reactivity in reactions of catalytic and induced decomposition, in reactions of bimolecular interaction with the nucleophilic reagents. The solution of these questions probably should be grounded on the interpretation of the experimental kinetic investigations from the point of view the reactivity indexes.

AFTERWORD

It is interesting to note the advisability of quantum–chemical investigations of the electron structure for some classes of organic peroxides under the aspect of the conformation and reactivity into reactions of thermal stability.

Kinetics and chemical mechanisms of the peroxides thermolysis reactions depend a great extent on the peroxides structure, on the decomposition medium, on the complex–formation, on the "cell" effect, on the viscosity and *ect.* [190–192].

Some interest represents the analysis of kinetic data of peroxides thermal decomposition in connection with their structure. The electron effects of the substitutes as a rule have an inessential influence on the rate of the alkylperoxides thermal decomposition [193–194]. The presence of α–C–H bonds into the peroxides leads to the formation of molecular hydrogen [195–197]; most probably, this is caused by the peroxides conformation and such chemical mechanism can be foreseen by supposed by us approach. The presence of double bond in the α–position to $-O$–$O-$ can lead to the α–decomposition of C–O bond [198–200]. Endoperoxides of anthracene and its derivatives can be easy decomposed with the formation of oxygen. The all of these chemical mechanisms can be forecasted based on the quantum–chemical calculations.

Dioxiranes are interesting as respects to their conformation [201–206]. Taking into account the complication of chemical mechanism of their thermal decomposition and considering also the following transformation of free radicals reactions, from our point of view the dioxiranes are quite adequate objects for the realization of our method.

Under the aspect of quantum–chemical calculations the dioxyethanes, which form the decomposition products into the activated state [207–211] are very interesting. $\Delta H^{\neq}$ of these compounds is within the interval of 85–140 $kJ \cdot mole^{-1}$.

The decomposition of trioxides [212–215] proceeds as a rule in accordance with the homolytical chemical mechanism with the formation of $RO^{\cdot}$ and $ROO^{\cdot}$ radicals. Proposed by us calculation of their reactivity permits to easy forecast the following reactions. It can be supposed, that the electron effects of the trioxides have an insignificant influence on kinetic parameters of the reactions, since the interval of the activation energies is comparatively narrow *(20–24 kilocalories/mole).*

Ozonides are decomposed at lower activation energies *(8–15 kilocalories/mole).*

Acylperoxides are characterized with a low thermal stability and are decomposed into solutions in accordance with the homolytical chemical mechanism [216–217].

Chemical mechanism of peroxyesters and peroxycarbonates decomposition [218–223] is connected with their non–centrality and with the presence of acylic group; this causes the possibility of competition between the homolytic and heterolytic decomposition ways and also causes the re–grouping ways realization.

Diacylperoxides are decomposed in accordance with both chemical mechanisms (homolytical and heterolytical ones) [224–231].

The percentage of heterolytic processes is increased at the stabilization increasing of the negative charge appearing into transition state [232]. Chemical mechanism of these compounds homolysis is represented in the above mentioned *ref.* [232].

So, the organic peroxides thermolysis into solution depending on the structure of decomposing compounds and the solvent's nature is realized both in accordance with the homolytical chemical mechanism with breaking of $-O-O-$ or $-CO-X-$ bonds and in accordance with the non–radical chemical mechanism including the coordinated re–grouping.

Let's note, that the proposed by us variant of the reactivity estimation permits to estimate also the kinetic parameters of the homolysis. Conformational analysis of the molecules of peroxides can a great extent forecast both the values of the kinetic parameters and the probable chemical mechanism of their decomposition in gaseous phase and also into the non–polar solvents.

It is clear, that a series of difficulties is appeared if the nature of solvent has a great influence on the reaction of peroxide decomposition; the nature of solvent can increase the part of one or other chemical mechanism.

Starting from our approach to the peroxides decomposition reactions it can be done the estimations of the forming radicals reactivity, it can be forecasted their stability, the decomposition products in monomolecular and bimolecular reactions. Our approach envisages the estimations of probability of heterolytic and induced peroxides decomposition.

From the other hand, the considered approach which is demonstrated on the reactions of peroxides can be also widened on any monomolecular reactions of the thermal decomposition of any organic molecules. Especially this is concerned to the reactions of the molecules, the properties of which can not be estimated by the additivity methods.

The lack of the proposed method for estimation of conformers' energy and the reactivity of molecules in the reactions of the thermal stability is its semi–quantitative estimation of kinetic parameters as a result of the semi–empirical quantum–chemical methods application.

The more perfect semi–empirical methods and improvement computers will be applied the more correct energetic characteristics of the non–valence atoms in molecules will be obtained; this practically will lead to more correct calculations of the molecules conformations and to more exact considering of the spatial effects influence on molecules reactivity into transformation reactions and on the estimation of the probabilities of these or others chemical mechanisms of the transformation.

Theoretically, at the present time the chemists can only dream of perfect calculations methods *ab inicio* which help investigate in detail a lot of chemical mechanisms of the chemical reactions.

REFERENCES

[1]   Silbert L. S., Witnauer L. P., Swern D., Ricciuti S. C. *X*–Ray Diffraction and Polarographic Study of *t*–Butyl Peresters and Diacyl Peroxides of Aliphatic Monobasic Acids // *«J. Amer. Chem. Soc.»*. – 1959. – v. 81. – № 17. – p. p. 3244–3250.

[2]   Liebiediev V. A., Drozdov Yu. N., Kuz'min E. A *and ect.* Specification of Crystalline Structure of Triphenyl(triphenylsililperoxy)silane *(C₆H₅)₃OO(C₆H₅)₃*: // *«Uspiekhi Khimiji»*. – 1981. – № 1. – p. p. 192–194 *(in Russian)*.

[3]   Karch N. J., Koh E. T., Whitsel B. L. and Bride M. J. An X–Ray and Electron Paramagnetic Resonance Structural Investigation of Oxygen Discrimination During the Collapse of Methyl–Benzoyl Oxy Radical Pairs in Crystalline Acetyl Benzoyl Peroxide // *«J. Amer. Chem. Soc.»*. – 1957. – v. 97. – № 23. – p. p. 6729–6743.

[4]   Kass J. D., Oberhammer H., Brandes D., Blaschette A. Electron Diffraction Investigation of the Molecular Structures of *di–t*–Butylperoxide and Bis(trimethylsilyl)–Peroxide // *«J. Mol. Struct.»*. – 1977. – v. 40. – № 1. – p. p. 65–75.

[5]   Liebiediev V. A., Drozdov Yu. N., Kuz'min E. A. *and ect.* Crystallic Structure of Triphenyl(triphenylgermylperoxy)silane // *«Doklady Akademiji Nauk SSSR»*. – 1979. – v. 246. – p. p. 601–605 *(in Russian)*.

[6]   Shklover V. E., Timofieyieva T. V., Struchkov Yu. T. The Structure of Organic and Silica–Organic Peroxides // Book of Abstracts of the VII *All–Union Conference on Chemistry of Organic Peroxides.* – Volgograd. – 1980. – p. p. 221–222.

[7]   Le Fevre R. J. W., Sundaram A. The Conformations of Some Simple Carboxylic Esters as Solutes // *«J. Amer. Chem. Soc.»* – 1962. – v. 10. – p. p. 3904–3914.

[8]   Verderame F. D., Miller I. G. The Electric Moments of Organic Peroxides. III. Peresters // *«J. Phys. Chem.»*. – 1962. – v. 66. – № 11. – p. p. 2185–2188.

[9]   Roges M. T., Campbell T. W. The Electric Moments of Some Aliphatic Peroxides, Disulfides and Diselenides // *«J. Amer. Chem. Soc.»*. – 1952. – v. 74. – № 19. – p. p. 4742–4743.

[10]   Yurzhenko T. I., Mamchur L. P., Vilenskaya M. R. Investigation of *IR*–Spectra of Hydroperoxies and Peresters // In Book: *«Uspiekhi Khimiji Organicheskikh Perekisnykh Soedinienij i Autookislieniya»*. – *M.:* *«Khimiya»*. – 1969. – p. p. 429–435 *(in Russian)*.

[11]   Anisimov Yu. N., Ivanchiov S. S., Yurzhenko A. I. Quantitative Determination of Diacyllic Peroxides by *IR*–Spectroscopy Method // *«Zhurn. Analitich. Khimiji»*. – 1966. – v. 21. – p. p. 113–116 (in Russian).

[12]   Kaldor U., Shavitt J. LCAO–SCF Computations for Hydrogen Peroxide // *«J. Chem. Phys.»*. – 1966. – v. 44. – № 5. – p. p. 1823–1829.

[13]   Dunning T. H., Winter H. W. Theoretical Determination of the Barriers to Internal Rotation in Hydrogen Peroxide // *«J. Chem. Phys.»*. – 1975. – v. 63. – № 5. – p. p. 1847–1855.

[14]   Pakiari A. H., Lennett J. W. Applications of a Simple Molecular Wave–Function. Part 9. Floating Spherical Gaussian Orbital Calculation for Hydrogen Peroxide // *«J. Chem. Soc. Faraday Trans.»*. – 1975. – Part 2. – v. 71. – № 9. – p. p. 1590–1594.

[15]   Semkov A. M., Linnett J. W. Applications of a Simple Molecular Wave–Function. Part 10. Analysis of the Rotational Barrier in Hydrogen Peroxide // *«J. Chem. Soc. Faraday Trans.»*. – 1975. – Part 2. – v. 71. – № 9. – p. p. 1595–1602.

[16]   Penney W. G., Sutherland G. B. The Theory of the Structure of Hydrogen Peroxide and Hydrazine // *«J. Chem. Phys.»*. – 1934. – v. 2. – № 8. – p. p. 492–498.

[17]   Giguere P. A., Schomaker V. An Electron Diffraction Study of Hydrogen Peroxide and Hydrazine // *«J. Amer. Chem. Soc.»*. – 1943. – v. 65. – № 10. – p. 2025.

[18]   Massay J. T., Bianco D. R. The Microwave Spectrum of Hydrogen Peroxide // *«J. Chem. Phys.»*. – 1954. – v. 22. – № 3. – p. p. 442–448.

[19] Lassettre E. N., Dean L. B. An electrostatic Theory of the Potential Barriers Hindering Rotation Around Single Bonds // *«J. Chem. Phys.».* – 1949. – v. 17. – № 3. – p. p. 317–332.

[20] Ionezawa T., Kato H., Iomamoto O. Orbital Investigation of the Electronic Structures of Alkyl Peroxides. II. Peroxyacids and Peresters // *«Bull. Chem. Soc., Jap.».* – 1967. – v. 40. – № 2. – p. p. 307–311.

[21] Sheppard N. Infrared Spectra and Assignment of the Fundamental Frequencies of the Alkyl Halides. I. *Tert*–Butyl Halides // *«Trans. Faraday Soc.».* – 1950. – v. 46. – p. p. 527–533.

[22] Kovner M. A., Koryakin A. V., Efimov A. P. Characteristic Frequencies of the Hydroperoxide Group–*COOH* // *«Optika i Spektroskopiya».* – 1960. – v. 8. – № 1. – p. p. 128–130 *(in Russian).*

[23] Anisimov Yu. N., Ivanchiov S. S., Yurzhenko A. I. Studies of Acyl Peroxides by *IR*–Spectroscopy Method // *«Tieoret. i Ekseriment. Khimiya».* – 1969. – v. 5. – № 4. – p. p. 518–522 *(in Russian).*

[24] Ivanchiov S. S., Anisimov Yu. N., Yurzhenko A. I. Applying the Method of *IR*–Spectroscopy for Studies of Diacylperoxides and Peresters // In Book: *«Uspiekhi Khimiji Organicheskikh Perekisnykh Soedinienij i Autookislieniya».* / *M.: «Khimiya».* – 1963. – p. p. 423–429 *(in Russian).*

[25] Midzusiama S. The Structure of Molecules and Internal Rotation. // *M.: «Inostr. Literat.».* – 1971. – 318 p.

[26] Turovskyj N. A., Turovskyj A. A., Rybachenko V. I. About Structure of *Tert.*–Butylperbenzoate // *«Ukr. Khim. Zhurn».* – 1980. – v. 46. – iss. I. – p. p. 99–102 *(in Russian).*

[27] Turovskyj N. A., Turovskyj A. A., Dzumedzej N. V. Conformational Analysis of *Tert.*–Butyl Peresters of Organic Acids // *«Zhurn. Struktur. Khimiji».* – 1977. – v. 18. – № 1. – p. p. 197–198 *(in Russian).*

[28] Sverdlov L. M., Kovner M. A., Kraynov E. P. Oscillation Spectra of Multiatomic Molecules // *M.: «Nauka».* – 1970. – 557 p. *(in Russian).*

[29] Ivanchiov S. S. Physical–Chemical Investigations of Organic Peroxide Compounds as Initiators of Radical Polymerization // *Abstract of the D. Sc. Thesis.* – Odessa. –. 1968. – 39 p. *(in Russian).*

[30] Turovskyj A. A., Turovskyj N. A., Rybachenko V. I., Chekushyn S. I. *IR*–Spectra and Composition of *Tert.*–alkylperoxymethyldialkylamines // *«Ukr. Khim. Zhurn».* – 1975. – v. 41. – № 5. – p. p. 634–636 *(in Russian).*

[31] Kucher R. V., Turovskyj A. A., Turovskyj N. A., Dzumedzej N. V., Dmytruk A. F. About Conformations of Some Asymmetric Alkyl

Peroxides // *«Doklady AN USSR»*. – 1973. – v. 210. – № 3. – p. p. 598–600 *(in Russian)*.

[32] Gribov L. A., Karyakin A. V. Intensities of *IR–* Adsorption Bands for Hydroperoxide Group // *«Optika i Spectroscopiya»*. – 1960. – v. 9. – № 5. – p. p. 666–668 *(in Russian)*.

[33] Orvill–Thomas Internal Rotation of Molecules // *M.: «Myr»*. – 1977. – 510 p. *(in Russian)*.

[34] Lide D. R. Jr. Structure of the Isobutane Molecule; Change of Dipole Moment on Isotopic Substitution // *«J. Chem. Phys.»*. – 1960. – v. 33. – № 5. – p. p. 1519–1522.

[35] Nishikawa T., Itoh T., Shinoda K. Molecular Structure of Methyl–Amine From Its Microwave Spectrum // *«J. Chem. Phys.»*. – 1955. – v. 23. – № 9. – p. p. 1735–1736.

[36] Lide D. R. Jr. Structure of the Methyl–Amine Molecule. I. Microwave Spectrum of CD_3ND_2 // *«J. Chem. Phys.»*. – 1957. – v. 27. – № 2. – p. p. 343–352.

[37] Davis M. J., Muecke A Molecular Structure Study of Cyclopentene // *«J. Chem. Phys.»*. – 1970. – v. 74. – № 5. – p. p. 1104–1108.

[38] Chiang J. F., Bauer S. H. The Molecular Structure of Cyclohexene // *«J. Amer. Soc.»*. – 1969. – v. 91. – № 8. – p. p. 1898–1901.

[39] Turovskj A. A., Turovskj N. A. About Conformational Analysis of Some Dialkyl Peroxy Compounds // *«Zhurn. Strukturn. Khim.»*. – 1979. – v. 20. – № 5. – p. p. 937–939 *(in Russian)*.

[40] Turovskj A. A., Petrenko V. V., Turovskj N. A., Litkoviets A. K. About Thermal Decomposition and Structure of Silicon–Containing Peroxides // *«Zhurn. Org. Khimiji»*. – 1977. – v. 47. – № 11. – p. p. 2556–2560 *(in Russian)*.

[41] Turovskj A. A., Turovskj N. A., Orlova L. V. About Structure of Some Alkyl Peroxides and *Tert.–*butyl Peresters of Organic Acids // *«Zhurn. Physich. Khimiji»*. – 1974. – v. 48. – iss. 10. – p. p. 2569–2571 *(in Russian)*.

[42] Kruglyak Yu. A., Dyadusha G. G., Kuprievich V. A., Podolska L. M., Kogan G. N. The Calculated Methods of Electron Structure and Spectra of Molecules // *K.: «Naukova dumka»*. – 1969. – 307 p. *(in Russian)*.

[43] Klopman G. Reactivity and Reactions Ways // *M.: «Myr»*. – 1977. – 383 p. *(in Russian)*.

[44] Iliel E., Allindzher N., Ential S., Morrison G. Conformational Analysis // *M.: «Myr»*. – 1969. – 592 p. *(in Russian)*.

[45] Bellami L. New Data upon *IR*–Spectra of Complicated Molecules // *M.: «Myr»*. – 1971. – 318 p. *(in Russian)*.

[46] Dashevsky V. T. Conformation of Organic Molecules // *M.: «Khimiya»*. – 1974. – 432 p. *(in Russian)*.

[47] Gertsberg G. Electron Spectra and Structure of Multiatomic Molecules // *M.: «Myr»*. – 1969. – 772 p. *(in Russian)*.

[48] Pople J. A., Beveridge D. L. Approximate Molecular Orbital Theory // *«Mc. Graw–Hill», N. Y.* – 1970. – 85 p.

[49] Ohkubo K., Kitagawa F. Estimation of the Structures and Electronic States of Radicals Using the *INDO–SC^F* MO Method. I. Alkoxyl Radicals // *«Bull. Chem. Soc.»*. – *Jap.* – 1974. – v. 47. – № 3, p. p. 739–740.

[50] Yarkony D. R., Shaefer H. F., Rothenberg S. X^3Ar, a^1E, $b'A_1$ Electronic States of Methylnitrene // *«J. Amer. Chem. Soc.»*. – 1974. – v. 96. – № 19. – p. p. 5974–5977.

[51] Ohubo K., Fujita T., Sato H. Conformations and Electronic Properties of Alkyl Peroxides, Alkylperoxy Radicals and Alkoxy Radicals // *«Mol. Struct.»*. – 1977. – v. 36. – № 1. – p. p. 101–110.

[52] Pshezhetskyj S. Ya., Kotov A. G., Milingu V. N. and other *EPR* of Free Radicals in Radiation Chemistry // *M.: «Myr»*. – 1972. – 460 p. *(in Russian)*.

[53] Osipov O. A., Minkin V. I., Garnovsky A. D. Hand–Book on Dipole Moments // *M.: «Vysshaya Shkola»*. – 1971. – 416 p.

[54] Biezuglyj V. D. Polarography in Chemistry and Technology of Polymers // *L.: «Khimiya»*. – 1968. – 231 p. *(in Russian)*.

[55] Swern D., Silbert L. S. Studies In The Structure of Organic Peroxides // *«Anal. Chem.»*. – 1963. – v. 35. – p. p. 880–885.

[56] Dzumedzej N. V., Turovskyj A. A., Turovskyj N. A. Polarographic Investigation of Some Organic Peroxides // *«Doklady AN USSR»*. – 1975. – Ser. *B*. – № 9. – p. p. 810–812 *(in Russian)*.

[57] Kucher R. V., Turovskyj A. A., Turovskyj N. A. // *Book of Abstracts on VI All–Union Meeting on Quantum Chemistry.* – Kisheniev. – 1975. – p. 146 *(in Russian)*.

[58] Kucher R. V., Turovskyj A. A., Turovskyj N. A. Quantum–Chemical Interpretation of Peroxides Radical Decomposition // *«Doklady AN USSR»*. – 1976. – v. 227. – № 3. – p. p. 611–613 *(in Russian)*.

[59] Turovskyj A. A., Kucher R. V., Turovskyj N. A., Lobanov V. V. About Quantum–Chemical Calculation of Decomposition Reaction of H_2O_2

and CH_3OOCH_3 In Gaseous Phase // *«Theor. i Ekspieriem. Khimiya».* – 1976. – v. 12. – № 5. – p. p. 606–611 *(in Russian).*

[60] Turovskyj A. A., Turovskyj N. A., Dolgushin M. D. Non–empirical Calculation of Radical Decomposition of Hydrogen Peroxide // *«Doklady AN USSR».* – 1976. – Ser. B. – № 8. – p. p. 724–725 *(in Russian).*

[61] Vishinskyj N. N., Kokorev V. N., Kulikova G. P. Quantum–Chemical Calculation of Dissociation Energy for Homonuclear Bonds in Organic Peroxides and Hydrozines // *«Doklady AN USSR».* – 1980. – v. 250. – p. p. 1140–1144 *(in Russian).*

[62] Litinsky A. O., Shrejbert A. I, Shyfrovych E. I., Baliavichus L.–M. Z., Bolozhyn A. V. Electron Structure, Stability and Reactivity of Acyl Peroxides // *«Theor. i Ekspieriem. Khimiya».* – 1973. – p. p. 692–696 *(in Russian).*

[63] Christopher G. Structure and Conformation in Molecular Peroxides // *«J. Mol. Struct.»* – 1980. – v. 67. – p. p. 35–44.

[64] Osamu Kikuchi Molecular Orbitals and $O–O$ Cleavage of Peroxides // *«J. Org. Chem.».* – 1979. – v. 37. – № 8. – p. p. 678–688.

[65] Hunt R. H., Leacock R. A., Peters C. W., Hecht K. T. Internal–Rotation in Hydrogen Peroxide. The Far–Infrared Spectrum and the Determination of the Hindering Potential // *«J. Chem. Phys.».* – 1965. – v. 42. – № 6. – p. p. 1931–1946.

[66] Ohkubo K., Kitagawa F., Nippon Kagaku Kaishi Theoretical Investigation of Geometry and Electron Properties of Alkoxy– and Alkylperoxy Radicals and Hydroperoxides of Alkyls // *«J. Chem. Soc., Jap.».* – № 11. – p. p. 2147–2151.

[67] Kondratyev V. N. Constants of Gaseous Reactions Rate // *M.: «Nauka».* – 1971. – 351 p. *(in Russian).*

[68] Dolgushin M. D. Non–empirical Calculations of Electron Structures of Molecules into Basis of the Gaussian's Functions // *K., preprint / Institute of Theoretical Physics: P. K.* – 1972. – 20 p. *(in Russian).*

[69] Dyuar M. Theory of Molecular Orbitals in Organic Chemistry // *M.: «Vysshaya Shkola».* – 1978. – 590 p. *(in Russian).*

[70] Davtyan O. K. Quantum Chemistry // *M.: «Vysshaya Shkola».* – 1962. – 781 p. *(in Russian).*

[71] Turovsky N. A. Investigation of Structure and Reactive Ability of the Peresters and Some Dialkylperoxides via Reactions of the Thermal Decomposition // *Ph. D. Thesis. – Donetsk.* – 1978. – 96 p. *(in Russian).*

[72] Nicholson A. E., Norrish R. G. Polymerization of Styrene at High Pressures Using the Sector Technique // *«Discuss. Faraday Soc.».* – 1956. – v. 22. – p. p. 104–113.

[73] Kucher R. V., Turovsky A. A., Turovsky N. A. About Connection Between the Reactive Ability of Organic Peroxides and Their Structure // *«Doklady AS USSR».* – 1978. – vol. 242. – № 5. – p. p. 1093 – 1095 *(in Russian).*

[74] Terman L. M. Chemistry of Oxygen–Containing Radicals // *«Uspiekhi Khimiji».* – 1965. – vol. 34. – iss. 3. – p. p. 434–453 *(in Russian).*

[75] Turovsky A. A., Kucher R. V. About Correlations Between Free Energies of Activated Bonds Dissociation and Free Energies of Starting Reagents Bonds Dissociation for Monomolecular Decomposition Reactions // *«Zhurnal Physicheskoj Khimiji».* – 1970. – vol. 44. – iss. 1. – p. p. 221 – 223 *(in Russian).*

[76] Hawkins E. G. Organic Peroxides // *«London».* – 1961. – 434 p.

[77] Nazin G. M. Preexponential Factor of Gaseous Monomolecular Reactions // *«Uspiekhi Khimiji».* – 1972. – vol. 41. – iss. 9. – p. p. 1537–1565 *(in Russian).*

[78] Benson S. Thermochemical Kinetics // *M.: «Myr».* – 1971. – 308 p. *(in Russian).*

[79] Serdyuk A. I., Turyvskyj A. A., Kucher R. V., Tytov E. V., Bielobrov V. M. The Influence of Solvents on Reactivity of Hydroperoxide of *Tert.*–Butyl in Reaction with 1,2,5,6–Tetrahydrobenzaldehyde // *«Zhurn. Organich. Khim.»* – 1975. – v. 11. – № 1. – p. p. 12–15 *(in Russian).*

[80] Viedienieiev V. I., Kibalko A. A. Constants Rate of Monomolecular Reactions // *M.: «Nauka».* – 1972. – 164 p. *(in Russian).*

[81] Dienisov E. T. Constants Rate of Homolytical Liquid–Phase Reactions // *M.: «Nauka».* – 1971. – 711 p. *(in Russian).*

[82] Puchin V. A., Yurzhenko T. I., Boysan O. E., Aparovych L.M. Investigation of Thermal Decomposition of Peresteric Monomers and Polymers // In Book: *«Uspiekhi Khimiji Organicheskikh Perekisnykh Soedinienij i Autookislieniya».* – *M.: «Khimiya».* – 1969. – p. p. 202–208 *(in Russian).*

[83] Hiatt R. R. The Decomposition of Silyl Peroxides // *«Can. J. Chem.».* – 1964. – v. 42. – p. p. 985–989.

[84] Nonhibel D., Wolton J. Chemistry of Free Radicals // *M.: «Myr».* – 1977. – 606 p. *(in Russian).*

[85] Coulson C. A., Longuet–Higgins H. C. The electronic Structure of Conjugated Systems. I. General Theory // *«Proc. Roy. Soc. (London)»*. – 1947. – v. A. – № 191. – p. p. 39 – 60.

[86] Higasi K., Baba H., Rembaum A. Quantum Chemistry // *M.: «Myr»*. – 1967. – 379 p. *(in Russian)*.

[87] Bazilevskyj M. V. The Method of Molecular Orbitals and Reactivity of Organic Molecules // *M.: «Myr»*. – 1969. – 303 p. *(in Russian)*.

[88] Turovsky A. A., Kucher R. V. Structure and Reactivity of Unstable Radicals // *«Doklady AS USSR»*. – 1978. – vol. 9B. – p. p. 821 – 823 *(in Russian)*.

[89] Batsanov S. S., Zviagina R. A. Integrals of Overlapping and the Problem of Effective Charges // *Novosibirsk: «Nauka»*. – 1966. – 309 p. *(in Russian)*.

[90] Zhogolev D. A., Volkov V. B. Methods, Algorithms and Programs for Quantum–Chemical Calculations of Molecules // *K.: «Naukova Dumka»*. – 1976. – 212 p. *(in Russian)*.

[91] Kucher R. V., Turovsky A. A., Turovsky N. A., Dzumedzej N. V. Kinetics and Direction of Thermal Decomposition of *Tert*–alkyl*(aryl)*peroxycyclopentenes–2 and *Tert*–alkylperoxytetrahydrobenzoates // *«Ukr. Khim. Zhurnal»*. – 1976. – v. 42. – iss. 3. – p. p. 255–258 *(in Russian)*.

[92] Dzumedzej N. V., Kucher R. V., Turovsky A. A., Koshovsky B. I. Investigation of Kinetics of Thermal Decomposition of Nitrogen–Containing Peroxides with *Tert*–alkyl Radical // *«Ukr. Khim. Zhurn.»*. – 1973. – v. 39. – iss. 11. – p. p. 1142–1145 *(in Russian)*.

[93] Kucher R. V., Turovsky A. A., Dzumedzej N. V. About Chemical Mechanism of Thermal Decomposition of Dialkylaminemethyl–*tert*–alkylperoxides // *«Doklady AS USSR»*. – 1973. – vol. 6B. – p. p. 533 – 536 *(in Russian)*.

[94] Turovsky A. A., Kucher R. V., Serdyuk A. I., Batog A. E. Studies of Influence of Medium on Kinetics of Decomposition of Nitrogen–Containing Bifunctional Peroxides with *Tert*–Alkyl Radical // *«Doklady AS USSR»*. – 1971. – vol. 5B. – p. p. 438 – 441 *(in Russian)*.

[95] Turovsky A. A., Kucher R. V., Batog A. E. Synthesis and Investigation of Some Nitrogen–Containing Peroxides // *«Zhurn. Organich. Khim.»* – 1971. – v. 7. – № 6. – p. p. 1141–1143 *(in Russian)*.

[96] Kucher R. V., Turovsky A. A., Batog A. E. Investigation of Decomposition Kinetics of Some *Tert*–Alkyl Peroxides // *«Zhurn. Organich. Khim.»* – 1969. – v. 5. – № 6. – p. p. 600–604 *(in Russian)*.

[97] Kucher R. V., Turovsky A. A., Opejda Y. A., Batog A. E. Investigation of Decomposition Kinetics of Some Bifunctional *Tert*–Alkyl Peroxides // *«Doklady AS USSR»*. – 1969. – vol. 7B. – p. p. 635 – 639 *(in Russian)*.

[98] Turovsky A. A., Kucher R. V., Batog A. E., Kiryushyna N. P. Synthesis and Decomposition Kinetics of Some Bifunctional Nitrogen–Containing Peroxides // *«Zhurn. Organich. Khim.»* – 1970. – v. 6. – p. p. 1227–1230 *(in Russian)*.

[99] Turovsky A. A., Kucher R. V., Batog A. E., Nikolayevsky A. N. Studies of Decomposition Kinetics of Nitrogen–Containing Bifunctional Peroxides with *Tert*–Alkyl Radical // // *«Doklady AS USSR»*. – 1970. – vol. 115. – p. p. 1015 – 1018 *(in Russian)*.

[100] Serrej A. Handbook On Organic Reactions // *M.: «Khim. Lit»*. – 1962. – 160 p. *(in Russian)*.

[101] Yablokov V. A., Yablokova N. V., Tarabarina A. P. Thermal Decomposition of Organicsilicon Peroxides by $(C_6H_5)_3SiOOC(CH_3)_{3-n}(C_6H_5)_n$ structure // *«Zhurn. Organich. Khim.»* – 1972. – v. 42. – № 5. – p. p. 1051–1055 *(in Russian)*.

[102] Yablokov V. A., Yablokova N. V., Tarabarina A. P., Shemuranova N. N. Thermal Decomposition of Organicsilicon Peroxides by $(CH_3)_3SiOOR$ structure // *«Zhurn. Organich. Khim.»* – 1971. – v. 41. – № 5. – p. p. 1565–1568 *(in Russian)*.

[103] Sluchevskaya N. P., Yablokov V. A., Yablokova N. V., Aleksandrov Y. A. Thermal Decomposition of Polyperoxysilanes // *«Zhurn. Organich. Khim.»* – 1976. – v. 46. – № 7. – p. p. 1540–1545 *(in Russian)*.

[104] Kucher R. V., Petrenko V. V., Turovsky A. A., Turovsky N. A., Litkovets A. K. About Structure and Homolysis of Organicsilicon Peroxides. The Structure and Reactivity of Organic Compounds // *In Book: «Collection of Scientific Works» / K.: «Naukova Dumka»*. – 1981. – 168 p. *(in Russian)*.

[105] Yablokov V. A., Sluchevskaya N. P., Yablokova N. V., Kalinin V.N., Izmajlov B. A. The Influence of Volumetric Substitutes on a Rate of Carbonylmethyl–Containing Silylperoxides Thermal Decomposition // *«Zhurn. Organich. Khim.»* – 1982. – v. 42. – № 7. – p. p. 893–896 *(in Russian)*.

[106] Petrenko V. V., Turovsky A. A., Litkovets A. K. Radical Decomposition of Silicon–Containing Peroxides // *«Ukr. Khim. Zhurn.»*. – 1979. – v. 45. – iss. 09. – p. p. 894–896 *(in Russian)*.

[107] Entelis S. G., Tigger R. P. Kinetics of Reactions in Liquid Phase // *M.: «Khimiya».* – 1973. – 416 p. *(in Russian).*

[108] Serbinov A. I. The Influence of Hydrogen Bond on Thermal Decomposition of Hydrogen Peroxide // *«Doklady AS USSR».* – 1982. – vol. 264. – № 5. – p. p. 1170 – 1172 *(in Russian).*

[109] Aleksandrov Y. A., Makin G. I., Liepikov V. E., Sokolova V. A., Mazanova L. M. Thermal Decomposition of Tri(*tert–*butylperoxy)gallium in Medium of Saturated and Unsaturated Organic Compounds // *«Zhurn. Organich. Khim.»* – 1982. – v. 42. – № 7. – p. p. 1575–1570 *(in Russian).*

[110] Shakhparonov M. I. Introduction into Molecular Theory of Solutions // *M.: «Goskhimizdat».* – 1958. – 296 p. *(in Russian).*

[111] Pimentel J., Mack–Klellan O. Hydrogen Bond // *M.: «Myr».* – 1964. – 462 p. *(in Russian).*

[112] Turovsky A. A., Kucher R. V., Titov E. V., Bielobrov V. M., Serdyuk A. I. About Hydrogen Bonds between Peroxide Compounds and *n–*Butyl alcohol // *«Zhurn. Physich. Khim.»* – 1973. – v. 47. – iss. 1. – p. p. 9–12 *(in Russian).*

[113] Turovsky A. A., Ustinova A. M., Serdyuk A. I., Batog A. E. Synthesis and Thin–Layer Chromatography of Organic Peroxide Compounds // *«Zhurn. Organich. Khim.»* – 1973. – v. 42. – iss. 5. – p. p. 1167–1171 *(in Russian).*

[114] Yogansen A. V., Rassadin B. V. Dependence of the Intensification and Displacement of *IR–*Bands *(OH)* on Energy of Hydrogen Bond // *«Zhurn. Prikladn. Spektrosk.»* – 1969. – v. 11. – iss. 5. – p. p. 827–829 *(in Russian).*

[115] Walling C., Gibian M. I. Hydrogen Abstraction Reactions by the Triplet States of Ketons // *«J. Amer. Chem. Soc.»* – 1965. – v. 87. – № 5. – p. p. 3361–3364.

[116] Bell E. R., Frederick F., Rust F. F., Vaughan W. E. Decomposition of *Di–t–*alkyl Peroxides. IV. Decomposition of Pure Liquid Peroxide // *«J. Amer. Chem. Soc.»* – 1950. – v. 72. – № 1. – p. p. 337–338.

[117] Lissi E. Photolysis of *Di–tert–*Butyl Peroxide in Solution // *«Can. J. Chem.»* – 1974. – v. 52. – № 13. – p. p. 2491–2492.

[118] Mc. Millan G., Wijnen M. H. J. Reactions of Alkoxy Radicals. V. Photolysis of *Di–t–*Butyl Peroxide // *«Can. J. Chem.»* – 1958. – v. 36. – № 9. – p. p. 1227–1232.

[119] Mc. Millan G. Photolysis of Diisopropyl Peroxide // *«J. Amer. Chem. Soc.»* – 1961. – v. 83. – № 14. – p. p. 3018–3023.

[120] Mc. Millan G. Photolysis of *Di–tert*–Butyl Peroxide – Azometane and *Di–tert*–Butyl Peroxide – Isobutane Mixtures // *«J. Amer. Chem. Soc.»* – 1960. – v. 82. – № 10. – p. p. 2422–2425.

[121] Norrish R. G., Searby M. H. The Photochemical Decomposition of Dicumyl Peroxide and Cumene Hydroperoxide in Solution // *«Proc. Roy. Soc.»* – 1956. – A 237. – № 1211. – p. p. 464–475.

[122] Ivanchiov S. S., Yurzhenko A. I., Lukovnikov A. F., Peredireyeva S. I. Studies of Photolytic Decomposition of Diacyl Peroxides by *EPR–* Method // *«Doklady AS USSR».* – 1966. – vol. 171. – № 4. – p. p. 894 – 897 *(in Russian).*

[123] Kaptein R., Hollander J. A.., Antheunis D., Osterhoff L. J. Chemically Induced Dynamic Nuclear Polarization. Triplet and Singlet State Photosensitization of Peroxide Decompositions // *«Chem. Communs.».* – 1970. – № 24. – p. p. 1687 – 1689.

[124] Fahrenholts S. R., Trozzolo A. M. The Alteration of Nuclear Magnetic Resonance Polarization in the Sensitized Photolists of Benzoyl Peroxide // *«J. Amer. Chem. Soc.»* – 1971. – v. 93. – № 1. – p. p. 251–253.

[125] Rykov S. V., Buchachenko A. L. Investigations of Nuclear Dynamic Polarization in Some Chemical Reactions // *«Doklady AS USSR».* – 1969. – vol. 185. – № 4. – p. p. 870 – 872 *(in Russian).*

[126] Krusic P. J., Kochi J. K. Electron Spin Resonanse of Primary Alkyl Radicals. Photolysis of Acyl Peroxides // *«J. Amer. Chem. Soc.»* – 1969. – v. 91. – № 14. – p. p. 3940–3942.

[127] Sheldon R. A., Kochi J. K. Photolysis of Peresters. Reaction of Alkoxy – Alkyl Radical Pares in Solution // *«J. Amer. Chem. Soc.»* – 1970. – v. 92. – № 17. – p. p. 5175–5186.

[128] Simpson W. H., Willer J. G. The Decomposition of Peroxy Esters. II. The Photolysis of Esters of Aliphatic Peroxycarboxylic Acids // *«J. Amer. Chem. Soc.»* – 1968. – v. 90. – № 15. – p. p. 4093–4095.

[129] Gray J. A., Style D. W. The Photolysis of Methyl Nitrile // *«Trans. Faraday Soc.».* – 1952. – v. 48. – p. p. 1137–1142.

[130] Petrenko V. V., Turovsky A. A., Litkovets' A. K. Photolysis of Organosilicon Peroxides in Liquid Phase // *«Zhurn. Organich. Khim.»* – 1981. – v. 51. – № 3. – p. p. 598–603 *(in Russian).*

[131] Rebbert R. E., Laidler K. J. Kinetics of the Decomposition of Diethyl Peroxide // *«J. Chem. Phys.».* – 1952. – v. 20. – № 4. – p. p. 574–577.

[132] Lukyanenko L. V., Petrenko V. V., Turovsky A. A., Turovsky N. A. Decomposition of Alkoxyradicals in Solid Phase at Low Temperatures

// *«Zhurn. Organich. Khim.»* – 1982. – v. 18. – p. p. 705–708 *(in Russian)*.

[133] Brinton R. K., Volman D. H. Decomposition of *Di–t–*Butyl Peroxide and Kinetics of the Gas Phase Reaction of *t–*Butoxy Radicals in the Presence of Ethylamine // *«J. Chem. Phys.»*. – 1952. – v. 20. – № 1. – p. p. 25–28.

[134] Gray P., Rathbone P., Williams A. Alicyclic Alkoxyl Radicals in the Thermal Decomposition of Cyclohexyl and 1–Methylcyclohexyl Nitrate // *«J. Amer. Chem. Soc.»* – 1961. – v. 90. – № 15. – p. p. 2620–2629.

[135] Piette L. H., Landgraf W. C. Steady–State Studies by *EPR* on the Photodissociation of Some Alkyl Hydroperoxides // *«J. Chem. Phys.»*. – 1960. – v. 32. – № 4. – p. p. 1107–1111.

[136] Ingold K. U., Morton J. R. Electron Spin Resonance Spectra of Organic Oxy Radicals // *«J. Amer. Chem. Soc.»* – 1964. – v. 86. – № 15. – p. p. 3400–3402.

[137] Ivanchiov S. S., Yurzhenko A. I., Lukovnikov A. F., Hak Yu. N., Kvasha S. M. Studies of Photolytic Decomposition of Diacyl Peroxides in Solid Phase and Kinetics of the Forming Radicals Decay // *«Theor. i Ekspieriem. Khimiya»*. – 1968. – v. 4. – iss. 6. – p. p. 780–787 *(in Russian)*.

[138] Zubkov A. V., Koritsky A. T., Lebediev Ya. S. Radicals's Pairs at the Photolysis of Peroxydicarbonic Acids // *«Doklady AS USSR»*. – 1969. – vol. 185. – № 6. – p. p. 1312 – 1315 *(in Russian)*.

[139] Kurita Y. Electron Spin Resonance Study of Radical Pairs Trapped in Ireadiated Single Crystals of Dimethylglyoxime at Liquidnitrogen Tempertaure // *«J. Chem. Phys.»*. – 1964. – v. 41. – № 12. – p. p. 3926–3927.

[140] Liebiediev L. S. Kinetics of Bimolecular Reactions in Condensed Phase. Calculation of Rate Based on Cell Model // *«Kinetika i Kataliz»*. – 1965. – vol. 8. – p. p. 245 – 249 *(in Russian)*.

[141] Lukyanenko L. V., Turovsky A. A., Kucher R. V. Studies of *UV–*Decomposition of *Tert–*Butyl Peroxide in Solid Phase // *«Zhurn. Organich. Khim.»* – 1982. – v. 12. – № 5. – p. p. 929–933 *(in Russian)*.

[142] Grinberg O. Ya., Dubinsky A. A., Liebiediev Ya. S. Separation of the Overlapping *EPR* Spectra by Differential Saturation Method // *«Doklady AS USSR»*. – 1971. – vol. 196. – № 6. – p. p. 627 – 629 *(in Russian)*.

[143] Kucher R. V., Turovsky A. A., Lukyanenko L. V., Dzumedzej N. V. Photolytic Decomposition of Nitrogen–Containing Peroxides with *Tert–*

alkyl Group in Solid Phase // *«Theor. i Ekspieriem. Khimiya».* – 1976. – v. 12. – № 1. – p. p. 41–47 *(in Russian)*.

[144] Fiel' N. S., Dolin I. P., Zolotariovsky V. I. Impulse radiolysis of formamides // *«Khimiya Vysokikh Energij».* – 1967. – v. 1. – p. p. 154–156.

[145] Leningston R., Zeldes H. Paramagnetic Resonanse Study of Liquids During Photolysis. V. Acid Amides. // *«J. Chem. Phys.».* – 1967. – v. 47. – p. p. 4173–4179.

[146] Backsendail J. K. Electron in Solution // *«Zhurn. Vsiesoyuzn. Obszch. im. D. I. Mendielieyeva».* – 1966. – v. 11. – № 2. – p. p. 168–177.

[147] Hatton J. V., Richards R. E. Solvent Effects in *NMR* Spectra of Amide Solutions // *«Mol. Phys.».* – 1962. – v. 5. – p. p. 139–152.

[148] Yonesawe T., Noda J., Kawamura T. Electron Spin Resonance Study of Radicals Formed by Abstracting Hydrogen From Amides // *«Bull. Chem. Soc. Jap.»* – 1969. – v. 42. – № 3. – p. p. 650–657.

[149] Bewington J. C. Radical Polymerization // *«N.–Y. Academic Press».* – 1961. – 112 p.

[150] Buchachenko A. L., Wassermann A. M. Stable Radicals. Electron Structure, Reactivity and Application // *M.: «Khimiya».* – 1973. – 408 p. *(in Russian)*.

[151] Mikhaylov I. I., Liebiediev L. S., Buben N. L. Recombination of Radicals in Solid Organic Substances // *«Kinetika i Kataliz».* – 1962. – v. 3. – p. p. 314–321 *(in Russian)*.

[152] Serheew H. V., Gurman V. J., Papisove V. U., Jacovenko R. J. Intersimposium on Free Radicals // Upsala, Sweden. – 1961. – v. 47.

[153] Buben N. L., Molin Yu. N. *EPR* Spectrum of Cyclohexyl Radical at Radiolysis of Cyclohexene in Gas–Crystalline State // *«Doklady AS USSR».* – 1963. – vol. 152. – p. p. 352 – 355 *(in Russian)*.

[154] Lukyanenko L. V., Turovsky A. A., Turovsky N. A. The Peculiarities of the Non–Symmetrical Acylperoxides Photodecomposition in Solid Phase // *«Zhurn. Organich. Khim.»* – 1979. – v. 49. – № 6. – p. p. 1398–1405 *(in Russian)*.

[155] Turovsky A. A., Petrenko V. V., Lukyanenko L. V., Litkovets A. K., Nikolayevsky A. N. Photolysis of Silicon–Containing Peroxides in Solid Phase // *«Deponent VINITI».* – 1976. – № 100–77 (presented by Donetsk State University; in Russian).

[156] Mc. Millan G. R. Excicted Alkoxy Radicals In The Photolysis Of Dialkyl Peroxides // *«J. Amer. Chem. Soc.»* – 1962. – v. 84. – № 13. – p. p. 2514–2519.

[157] Tyudesh F. T. Consideration Of Radical Polymerization Kinetics Based On The Hypothesis Of Hot Radicals // *M.: «Myr»*. – 1968. – 148 p. *(in Russian)*.

[158] Bogatyryova A. I., Buchachenko A. L. Reactions Of Electron–Activated Radicals And Quenchering The Activated States By Radicals // *«Uspiekhi Khimiji»* – 1975. – v. 44. – № 12. – p. p. 2171–2203 *(in Russian)*.

[159] Turro N. Molecular Photochemistry // *M.: «Myr»*. – 1967. – 328 p. *(in Russian)*.

[160] Walling C., Indictor N. Solvent Effects And Initiator Efficiency In The Benzoil Peroxide – Dimethylaniline System // *«J. Amer. Chem. Soc.»* – 1958. – v. 80. – № 21. – p. p. 5814–5818.

[161] Horner L., Schwenk E. The Polymerization Of Vinyl Compounds In The System: Diacyl Peroxide – *Tert*–Amine // *«Annual Reports»* – 1950. – v. 566. – p. p. 69–84.

[162] Horner L., Kirmse W., Liebig S. Studien Zum Ablauf Der Substitution. IX Zur Oxidation Entalkylierung Gemischt Substituierter *Tert*–Tiarer Amine Durch Dibenzoylperoxed // *«Annual Chem. Reports»* – 1955. – v. 597. – № 1. – p. p. 48–54.

[163] Kucher R. V., Lukyanenko L. V., Turovsky A. A. Peculiarities Of 1,1–Diphenylpikrylhydrazyne Interaction With Peroxides Containing Of *Tert*–Butyl Group // *«Zhurn. Organich. Khim.»* – 1981. – v. 17. – № 4. – p. p. 837–843 *(in Russian)*.

[164] Lee J. B., Uff B. C. Organic Reactions Insolving Electrophilic Oxygen // *«Guart. Rev.»*. – 1967. – v. 21. – № 4. – p. p. 429 – 457.

[165] Chaltykyan O. A., Akopyan C. A., Beyleryan N. M., Salukhanyan E. R. Studies Of The Kinetics Of Free Radicals Generation At The Reaction Of Benzoil Peroxide With Triethylamine In Benzene // *«Scientific Notes Of Erevan State University. Natural Sciences»*. – 1972. – № 2. – p. p. 40 – 46.

[166] Buchachenko A. L., Pobiedimsky D. G., Neyman M. B. About Chemical Mechanism Of Peroxides Reactions With The Aromatic Amines // *«Zhurn. Fizich. Khimiji»*. – 1968. – v. 42. – № 6. – p. p. 1436–1440.

[167] Razuvayev G. A., Terman L. M., Mikhotova L. N., Yanovsky D. M. Radical Reactions of Organic Peroxydicarbonates. III. An Interaction Of Dicyclohexylperoxydicarbonate With The Dimethylaniline // *«Zhurn. Organich. Khim.»* – 1965. – v. 1. – № 1. – p. p. 79–82 *(in Russian)*.

[168] Zupancic J., Horn K., Schuster G. Electron – Transfer Initiated Reactions Of Organic Peroxides With Olefins And Other Electron Donors // *«J. Amer. Chem. Soc.»* – 1980. – v. 102. – № 21. – p. p. 5279–5285.

[169] Alexandrov Yu. A., Horbatov V. V., Yablokova N. V. An Interaction Of Hetero–Organic Peroxides Of The Silicon Sub–Group With The Amines // *«Doklady AS USSR».* – 1980. – vol. 250. – p. p. 623 – 626 *(in Russian).*

[170] Pobiedimsky D. G., Razumovsky S. D. About Reaction Of Aromatic Amines With The Ozonides Of Olefines // *«Izv. AS USSR».* – 1970. – № 3. – p. p. 602 – 605 *(in Russian).*

[171] Mulliken R. S. Molecular Compounds And Their Spectra // *«J. Amer. Chem. Soc.»* – 1952. – v. 74. – № 3. – p. p. 811–824.

[172] Horner L., Schwenk E. The Acceleration of Peroxide Decomposition By Amines // *«Angew. Chem.»* – 1949. – v. 61. – p. p. 411–412.

[173] Horner L. Zur Umsetzung Von Diacylperoxyden Mit *Tert*–Tiaren Aminen // *«J. Polymer Sci.»* – 1955. – v. 18. – № 89. – p. p. 438–439.

[174] Markaryan Sh. P., Beyperyan N. M. Chemical Polarization on Nuclears In Reactions Of Benzoil Peroxides With Trimethylamine // *«Armenian Chem. Jour.»* – – 1981. – v. 17. – p. p. 424–427.

[175] Sogomonyan B. M., Beyleryan N. M. Kinetic Regularity Of The Chains Initiation Under Oxidizing Of Triethanolamine By Benzoil Peroxide In Different Solvents // *«Armenian Chem. Jour.»* – – 1980. – v. 33. – p. p. 880–884.

[176] Beyleryan N. M. About Correlation Between The Ionization Potential Of Amines And Rate Of Their Oxidizing By Peroxides // *«Scientific Notes Of Erevan State University. Natural Sciences».* – 1971. – № 1. – p. p. 128 – 145.

[177] Chaltykyan O. A., Atanasyan E. N., Sarkisyan A. S., Marmaryan G. A., Haybakyan D. M. Kinetics Of Interaction Of Benzoil Peroxide With Secondary Aliphatic Amines // *«Zhurn. Fizich. Khimiji».* – 1958. – v. 32. – № 11. – p. p. 2601–2607.

[178] Pobiedimsky D. G. Kinetics And Chemical Mechanism Of The Peroxide Compounds With The Phosphites, Sulphides And Aromatic Amines Interaction // *«Uspiekhi Khimiji»* – 1971. – v. 40. – № 2. – p. p. 254–275 *(in Russian).*

[179] Yarmolyuk B. M., Polumbrik O. K., Dvorko H. F. Kinetics Of Reaction Of Benzoil Peroxide With Triphenylverdazyl Radical // *«Nieftiekhimiya»* – 1973. – v. 13. – № 5. – p. p. 719–722 *(in Russian).*

[180] Yarmolyuk B. M., Polumbrik O. K., Dvorko H. F. About Regeneration Of Triphenylverdazyl In Reaction With Benzoil Peroxide In Alcohol // *«Kinetika i Kataliz»*. – 1973. – v. 14. – p. p. 1592–1593 *(in Russian)*.

[181] Pokhodenko V. D., Bielodied A. A., Koshechko V. G. Oxidizing–Reduction Reactions Of Free Radicals // *K.: «Naukova Dumka»*. – 1977. – 276 p. *(in Russian)*.

[182] Pokhodenko V. D. Phenoxyl Radicals // *K.: «Naukova Dumka»*. – 1969. – 196 p. *(in Russian)*.

[183] Turovsky A. A., Kucher R. V., Lukyanenko L. V. Kinetics Of *DPhPH* Interaction With Some Non–Symmetric Alkyl Peroxides And Peresters // *«Zhurn. Organich. Khim.»* – 1975. – v. 45. – № 5. – p. p. 1137–1140 *(in Russian)*.

[184] Petrenko V. V., Kucher R. V., Turovsky A. A., Lukyanenko L. V., Litkovets A. K., Andrianov A. A. About Chemical Mechanism Of Interaction Of Some Organic Silicon–Containing Peroxy Compounds With *DPhPH* // *«Doklady AS USSR»*. – 1979. – vol. 4B. – p. p. 284 – 287 *(in Russian)*.

[185] Turovsky A. A., Lukyanenko L. V., Petrenko V. V., Andrianov A. A., Kotlyarenko V. V., Litkovets A. K. Peculiarities Of *DPhPH* Interaction With Some Organic Silicon–Containing Peroxy Compounds // *«Deponent VINITI»*. – 1978. – № 1904–78 *(presented by Donetsk State University; in Russian)*.

[186] Reinhardt X. Solvents in Organic Chemistry // *L.: «Khimiya»*. – 1973. – 137 p. *(in Russian)*.

[187] Turkevich J., Lelwood P. Solid Free Radical As Catalyst For Ortho–Para Hydrogen Conversion // *«J. Amer. Chem. Soc.»* – 1941. – v. 63. – № 3. – p. p. 1077–1079.

[188] Antonovsky V. L., Bezborodov L. D., Yaselmann M. E. Studies Of The Initiating Ability Of Organic Peroxides By Inhibitors Method // *«Zhurn. Fizich. Khimiji»*. – 1964. – v. 43. – № 11. – p. p. 281–284.

[189] Turovsky A. A., Dzumedzej N. V., Turovsky N. A., Lukyanenko L. V. Reaction Of Interaction Of *Tert*–Butylperbenzoate With α–Naphthylamine // *«Deponent VINITI»*. – 1976. – № 3821–76 *(presented by Donetsk State University; in Russian)*.

[190] Dienisov E. T., Sarkisov O. M., Lichtenstein G. I. Chemical Kinetics // *M.: «Khimiya»*. – 2000 *(in Russian)*.

[191] Bazylievsky M. V., Faustov V. I. // *«Uspiekhi Khimiji»* – 1992. – v. 61. –p. 1185 *(in Russian)*.

[192] Antonovsky V. L., Zhulin V. M. // *«Kinetika i Kataliz»*. – 2003. – v. 44. – № 80 *(in Russian)*.

[193] Molyneux P. // *«Tetrahedron Lett.»*. – 1966. – vol. 22. – p. 2929.

[194] Richardson W. H., Yelvington M. B., Andrist A. H., Enley E. W., Smith R. S., Johnson T. D. // *«Journ. Org. Chem.»*. – 1973. – v. 38. – p. 4219.

[195] Hiatt R., Nair V. G. K. // *«Can. Journ. Chem.»*. – 1980. – v. 58. – p. 450.

[196] Hiatt R., Rahimi R. M. // *«Int. Journ. Chem. Kinet.»*. – 1978. – v. 10. – p. 185.

[197] Hiatt R. // *In: «Organic Free Radicals» (ASC Symp. Ser.)*. – Ed. W. A. Pryor. – *American Chemical Society, Washington, DC.* – 1978. – p. 63.

[198] Okubo M., Iba–Imai E., Mizuta H., Kawamoto T. // *«Bull. Chem. Soc. Jpn.»* – 1976. – vol. 49. – p. 1717.

[199] Turro N. J., Chow M.–F., Rigaudy J. // *«Journ. Amer. Chem. Soc.»*. – 1979. – v. 101. – p. 1300.

[200] Turro N. J., Chow M.–F. // *«Journ. Amer. Chem. Soc.»*. – 1981. – v. 103. – p. 7218.

[201] Adam W., Chan Y.–Y., Cremer D., Gauss J., Scheutzow, Schindler M. // *«Journ. Org. Chem.»*. – 1987. – v. 52. – p. 2800.

[202] Hull L. A., Budhai L. // *«Tetrahedron Lett.»*. – 1993. – vol. 34. – p. 5039.

[203] Khursan S. L. // *D. Sc. Thesis / Institute of Organic Chemistry / Russian Academy of Sciences / Upha.* – 1999.

[204] Khursan S. L., Grabovsky S. A., Kabal'nova N. N., Galkin E. G., Shereshoviets V. V. // *«Izviestiya Akademiji Nauk»*. – 2000. – p. 1344 *(in Russian)*.

[205] Minisci F., Zhao L., Fontana F., Bravo A. // *«Tetrahedron Lett.»*. – 1995. – vol. 36. – p. 1697.

[206] Kazakov D. V., Khusnullina D. R., Kabal'nova N. N., Khursan S. L., Shereshoviets V. V. // *«Izviestiya Akademiji Nauk»*. – 1997. – p. 1785 *(in Russian)*.

[207] Schuster G. R. // *«Acc. Chem. Res.»* – 1979. – vol. 12. – p. 366.

[208] Adam W. // *«Adv. Heterocycl. Chem.»* – 1977. – vol. 21. – p. 437.

[209] Adam W. // *In book: «The Chemistry of Peroxides» / Ed. S. Patai.* – Wiley, New York. – 1983. – p. 829.

[210] Vasilyev R. F. // *In book: «The Chemistry of Organic Peroxides» / Ed. S. Emmanuel.* – Volgograd Polytechnical Institute, Volgograd. – 1982. – p. 75.

[211] Sharipov G. L., Kazakov V. P., Tolstikov G. A. The Chemistry and Chemiluminescence of 1,2–Dioxyethanes // *M.: «Nauka»*. – 1990 *(in Russian)*.

[212] Khalizov A. F., Khursan S. L., Shereshoviets V. V. // *«Kinetika i Kataliz»*. – 1999. – vol. 40. – p. 216.

[213] Bartlett P. D., Lahav M. // *«Jsr. J. Chem.»* – 1972. – vol. 10. – p. 101.

[214] Khursan S. L., Shereshoviets V. V., Shishlov N. M., Khalizov A. F., Komissarov V. D. // *«React. Kinet. Catal. Lett.»* – 1994. – vol. 52. – p. 249.

[215] Kurz M. E., Pryor W. A. // *«J. Am. Chem. Soc.»* – 1978. – vol. 100. – p. 7953.

[216] Lefort D., Fossey J., Grusselle M., Nedelec J., Sorba J. // *«Tetrahedron Lett.»*. – 1985. – vol. 41. – p. 4273.

[217] Sorba J., Lefort D., Fossey J., Grisel F., Sanderson W. // *«J. Chem. Soc., Perkin Trans. 2»*. – 1992. – p. 1565.

[218] Ryuhardt H. // *«Uspiekhi Khimii»*. – 1968. – vol. 37. – p. 1402.

[219] Ryuhardt H. // *«Uspiekhi Khimii»*. – 1978. – vol. 47. – p. 2014.

[220] Yablokov V. A. // *«Uspiekhi Khimii»*. – 1980. – vol. 49. – p. 1711.

[221] Yablokov V. A. // In book: *«Chemistry of Organic Peroxides»*. / Ed. S. Emmanuel. – Volgograd Polytechnical Institute, Volgograd. – 1982. – p. 103.

[222] Criegee R. // *«Fortschr. Chem. Forsch.»* – 1950. – p. 508.

[223] Bartlett P. D. // In: *«Peroxide Reaction Mechanisms»* / Ed. J. Edwards. – Interscience, New York. – 1961. – p. 1.

[224] Fujimori K., Hirose Y., Oac S. // *«J. Chem. Soc., Perkin Trans. 2.»* – 1996. – p. 405.

[225] Fujimori K., Hirose Y., Oac S. // *«J. Chem. Soc., Perkin Trans. 2.»* – 1996. – p. 413.

[226] Nozaki K., Bartlett P. D. // *«J. Chem. Soc.»* – 1946. – vol. 68. – p. 1686.

[227] Pokidova T. S., Dienisov E. T. // *«Nieftiekhimiya»* – 1998. – vol. 38. – p. 269.

[228] Dienisov E. T. // *«Kinetika i Kataliz»* – 1999. – vol. 40. – p. 239.

[229] Stankievich A. I., Ziat'kov I. P., Lazarieva A. M., El'nitskiy A. P. // *«Zhurn. Organich. Khim.»* – 1980. – v. 16. – p. 1823 *(in Russian)*.

[230] Stankievich A. I. *«Kinetika i Kataliz»* – 1988. – vol. 29. – p. 684.

[231] Stankievich A. I. *«Kinetika i Kataliz»* – 1989. – vol. 30. – p. 781.

[232] Antonovskii V. L., Khursan S. L. Thermolysis of Organic Peroxides in Solution // *«Russian Chemical Reviews»* – 2003. – vol. 72 (11). – p. p. 939–963.

INDEX

A

acetone, 164, 169, 178, 179, 181, 182, 183, 199, 200, 203
acid, 29, 38, 51, 132, 147
activation energy, 52, 53, 54, 56, 59, 60, 70, 72, 77, 78, 79, 80, 86, 91, 94, 95, 96, 98, 111, 115, 117, 118, 119, 124, 132, 134, 145, 154, 155, 158, 167, 169, 189, 190, 192, 197, 198, 209, 210, 212, 214, 215, 216, 219
activation enthalpy, 80
activation entropy, 82, 83, 84, 85, 86, 89, 146, 147, 198, 209, 210
adsorption, 75, 160
alcohols, 42, 140
aldehydes, 29, 31, 138
aliphatic amines, 196
amine, 195, 196, 214, 216
amines, 140, 195, 196
amorphous phases, 161
atom potential, 30, 32
atomic orbitals, 31, 112, 116
axial position, 26

B

barriers, 30, 31, 32, 129
benzene, 84, 86, 128, 144, 147, 205

bonds, 7, 12, 13, 15, 16, 25, 34, 36, 40, 53, 56, 61, 67, 68, 69, 72, 75, 80, 82, 84, 86, 89, 90, 91, 93, 95, 104, 108, 121, 124, 138, 151, 152, 153, 154, 155, 156, 157, 160, 189, 191, 192, 199, 209, 210, 223, 224
butyl ether, 4

C

capillary, 50
carbon, 36, 37, 71, 74, 107, 148, 175, 180
carbon atoms, 74
carbon tetrachloride, 148
carbonyl groups, 160
cation, 196
CH3COOH, 81
charger distribution, x
chemical, ix, x, xi, 13, 29, 32, 42, 49, 52, 54, 63, 67, 68, 72, 75, 79, 98, 101, 103, 108, 111, 127, 138, 142, 151, 157, 158, 159, 161, 165, 169, 190, 193, 195, 196, 197, 206, 215, 216, 219, 220, 223, 224, 225
chemical bonds, 52, 72
chemical characteristics, 67, 101, 103
chemical interaction, 32
chemical properties, ix, 75, 158
chemical reactions, 109, 225
chemical reactivity, x

chemiluminescence, 135
chlorine, 29, 73
chloroanhydrides, 29
chlorobenzene, 132, 135, 136, 137, 139, 140, 142, 143, 145, 146, 148, 155, 162, 163, 164, 166, 168, 171, 173, 182, 198, 200, 201, 202, 203, 204, 208, 209, 210, 211, 212, 213, 214, 215
chloroform, 163, 169, 197, 198, 208
classification, xi, 68, 89, 90, 100, 102, 105, 221
CNDO method, 30
CO2, 157, 160, 181
collisions, 162, 190
compounds, x, 1, 3, 13, 14, 22, 26, 29, 30, 31, 42, 43, 49, 52, 55, 61, 75, 79, 80, 82, 86, 89, 91, 100, 110, 127, 131, 132, 134, 137, 138, 142, 144, 145, 148, 151, 153, 158, 159, 160, 166, 181, 182, 183, 187, 189, 191, 195, 197, 203, 208, 209, 215, 216, 224
condensation, 140
configuration, 4, 56, 187
conformational analysis, 3, 15, 21, 23, 24, 26, 32, 54, 91, 219
constant rate, 163
construction, 5, 13
consumption, 51, 144, 197, 198, 202, 210, 214
cooling, 161
Coulomb energy, 98
Coulomb interaction, 28
cycloalkenyl peroxides, 1

D

decay, 162, 163, 168, 169, 172, 173, 193
decomposition reactions, x, xi, 108, 156, 225
decoupling, 67
deficiency, 34
derivatives, 1, 2, 127, 128, 130, 131, 133, 147, 223
detachment, 108, 124, 132, 133, 134, 138, 139, 157, 160

deviation, 107, 135
dielectric constant, 29, 148
dielectric permeability, 145, 146, 148
differential equations, 165
diffraction, 3, 29
diffusion, 37, 219
dimensionality, 105
dimethylformamide, 51, 145, 146, 148, 167, 169, 203
dimethylsulfoxide, 203
dipole moment, 15, 16, 17, 18, 19, 20, 21, 23, 24, 25, 26, 27, 30, 32, 34, 38, 42, 43, 44, 45, 48, 49, 148, 204, 205
dipole moments, 17, 18, 20, 23, 24, 25, 26, 27, 30, 32, 34, 42, 43, 44, 48, 49, 148, 204
displacement, 7, 12, 13
disposition, 40
dissociation, 56, 60, 64, 79, 161, 187
distribution, x, 34, 42, 46, 49, 64, 66, 67, 68, 70, 71, 73, 111, 112, 191
disturbance energy, 109, 112, 117, 118, 119
divergence, 60, 83
DMFA, 207
double bonds, 132

E

electrolysis, 51
electron pairs, 75
Electron Paramagnetic Resonance (EPR), 161, 163, 166, 167, 168, 173, 174, 175, 176, 177, 180, 181, 182, 183, 196, 227, 231, 237, 238, 239
electronic structure, 74, 90, 148, 221
electroreduction, 51
empirical methods, x, 30, 34, 59, 225
endowments, 82, 83, 84, 111, 120
energetic characteristics, 225
energetic parameters, 95
entropy, 82, 84, 86, 89, 146, 148, 210
epoxycyclopentanone, 130
equilibrium, 50, 55, 56, 61, 62, 63, 67, 70, 71, 72, 74, 80, 81, 82, 83, 84, 86, 89, 91, 93, 95, 97, 98, 100

ethers, 1, 29, 31, 72
evidence, 31, 65, 141, 148, 159, 160, 175, 187, 189, 199, 214
experimental condition, 98
extinction, 159

F

flatness, 69
force, 5, 7, 12
force constants, 5
formaldehyde, 29, 140
formamide, 128, 167, 171
formation, 36, 43, 65, 69, 72, 77, 79, 80, 81, 91, 95, 100, 108, 124, 129, 130, 132, 138, 140, 142, 146, 148, 156, 157, 160, 163, 179, 182, 191, 192, 196, 199, 205, 206, 209, 215, 216, 223, 224
formula, x, 31, 54, 91, 95, 138
fragments, 12, 13, 74, 75, 182, 183
free activation energy, 80
free energy, 80
free radicals, ix, xi, 3, 55, 57, 63, 84, 111, 113, 115, 121, 122, 123, 124, 125, 128, 130, 138, 145, 155, 156, 161, 190, 193, 196, 215, 221, 223
free rotation, 82

G

gallium, 236
geometry, 3, 12, 13, 36, 38, 39, 40, 49, 56, 59, 63, 70
graph, 88
gravitation, 119
grouping, 224

H

halogen, 29, 30
HCC, 84
hexane, 163, 181, 182, 203, 204
homolytic, ix, xi, 3, 60, 70, 86, 91, 96, 151, 224

hydrocarbons, x, 29, 181, 182
hydrogen, 4, 8, 29, 36, 37, 59, 107, 108, 109, 110, 111, 112, 120, 121, 122, 124, 129, 130, 132, 133, 134, 138, 139, 140, 142, 144, 146, 148, 149, 156, 157, 160, 164, 167, 169, 176, 182, 183, 188, 191, 192, 209, 216, 220, 223
hydrogen atoms, 8, 192
hydrogen bonds, 146, 149
hydrogen peroxide, 4
hydroperoxides, 160
hydroxyl, 35, 50, 75
hyperfine interaction, 175

I

illumination, 38, 159, 163, 167, 170, 171, 177, 182, 189
inhibition, 135, 144
inhibitor, 130, 135, 137, 144, 187, 210, 211, 212
initial state, 83
initiation, ix, 187
interaction effect, 73, 112
interaction effects, 112
inversion, 175
ionization, 34, 75, 196
ionization potentials, 34, 196
IR spectra, 169
isobutane, 14
isobutylene, 164, 169, 178, 183
isolated molecule, 109, 116
isomerization, 107, 108, 109, 124, 125, 157, 220
isomers, 5, 219

K

kinetic curves, 163, 197, 198, 207
kinetic parameters, ix, 1, 87, 98, 100, 105, 135, 145, 158, 203, 210, 214, 219, 221, 224, 225

kinetics, ix, xi, 32, 54, 112, 122, 127, 135, 137, 138, 139, 145, 146, 148, 157, 159, 162, 169, 197, 203, 220

L

lead, ix, 4, 18, 43, 71, 116, 127, 148, 151, 158, 215, 223, 225
linear dependence, 50, 80, 95, 146, 171, 202
liquid phase, 98, 108, 159, 164, 186, 191, 193
liquids, 164, 219
localization, 30, 67, 69, 72, 109
logarithmic coordinates, 171
low temperatures, 175, 192

M

magnitude, 13, 56, 80, 96, 98, 103, 109, 110, 116, 156, 171, 187, 189, 216, 220
melting temperature, 200
mercury, 50, 51, 108, 171
methanol, 108
methyl group(s), 18, 25, 29, 84, 124, 167, 199
molecules, x, xi, 3, 5, 12, 13, 24, 28, 30, 31, 32, 57, 63, 65, 67, 69, 70, 73, 75, 76, 77, 79, 81, 82, 86, 108, 112, 116, 135, 138, 139, 159, 187, 189, 190, 191, 203, 220, 221, 224, 225
monomolecular reactions, 80, 189, 225

N

nitrobenzene, 145, 146, 148
nitrogen, 12, 13, 16, 17, 29, 40, 67, 84, 134, 135, 137, 138, 139, 140, 141, 142, 158, 161, 166, 167, 169, 173, 190, 191, 192, 197, 203, 209, 215, 216
NMR, 29, 199, 200, 203, 205, 239
Nuclear Magnetic Resonance, 237
nucleophiles, xi, 197

O

olefins, 108, 112
organic compounds, 31, 49
organic peroxides, ix, x, xi, 1, 2, 3, 32, 54, 55, 68, 75, 82, 90, 91, 96, 99, 105, 195, 219, 220, 221, 223, 224
oscillation, 7, 8, 12, 13, 81, 82, 83, 84, 86
oxygen, 7, 21, 24, 27, 28, 29, 30, 34, 36, 37, 38, 40, 43, 44, 48, 49, 54, 57, 58, 59, 67, 68, 69, 71, 72, 73, 75, 77, 78, 79, 91, 100, 109, 110, 112, 121, 122, 124, 129, 132, 145, 157, 161, 168, 215, 216, 220, 221, 223

P

penetrability, 203, 204
permeability, 146
phenoxyl radicals, 197
photolysis, xi, 159, 160, 163, 164, 165, 166, 168, 169, 172, 173, 174, 177, 178, 179, 181, 182, 187, 188, 192, 193
physical properties, 75
point of origin, 50
polar, 28, 29, 58, 69, 109, 118, 120, 148, 196, 203, 205, 224
polarity, 146, 199, 203, 206
polarizability, 69
polymerization, 1, 137
polymerization process, 137
primary products, ix, 121, 128, 132, 156, 157, 191
probability, 64, 75, 79, 89, 124, 187, 188, 190, 192, 215, 219, 225
prognosis, 153, 155
protons, 77, 174, 175, 177, 182, 183, 192, 199, 203
purity, 12
pyrolysis, 108

Q

qualitative concept, 70

quantitative estimation, 225
quantum yields, 187
quantum–chemical calculation, xi, 55, 138, 223

R

radiation, 231
radical formation, 221
radical mechanism, 195
radical pairs, 189
radical polymerization, ix
radical reactions, 206
reaction decomposition, 80
reaction order, 197, 203, 210
reaction rate, 187, 198, 202, 203, 204, 214
reagents, 28, 79, 80, 108, 116, 117, 118, 203, 204, 205, 207, 209, 210, 214, 221
recombination, 80, 140, 161, 165, 215
redistribution, 70
repulsion, 28, 29, 30, 32, 34, 61, 62, 102, 119, 120
resins, ix
resonance integrals, 29, 117
reverse reactions, 79
room temperature, 203
rotations, 89
rubbers, ix, 155

S

saturated hydrocarbons, 112
saturation, 163, 166
sensitivity, 70
silane, 227
silicon, 122, 182, 183, 203, 204, 206, 215, 216
skeleton, 4, 38, 43
solid matrix, 172, 175, 189
solid phase, 160, 162, 164, 165, 166, 168, 169, 179, 187, 188, 191, 192, 193
solid state, 159
solution, ix, x, xi, 54, 108, 145, 161, 163, 185, 199, 221, 224

solvation, 96, 135, 146, 148, 158, 166, 198
solvents, 108, 128, 135, 137, 141, 144, 145, 147, 158, 196, 199, 203, 207, 224
spatial location, 42
spectroscopy, 4, 29, 54
spin, 35, 36, 37, 43, 44, 48, 54, 75, 82, 110, 111, 112, 119, 129, 132, 156, 163, 171, 176, 177, 182
stability, 4, 29, 31, 54, 75, 78, 107, 108, 119, 225
styrene, 137, 139, 142, 145, 148, 212
styrene polymerization, 139, 142
substitutes, 1, 12, 34, 43, 46, 49, 55, 69, 71, 72, 73, 74, 89, 100, 110, 111, 129, 132, 151, 152, 158, 192, 219, 223
substitution, 69, 109, 121
sulphur, 21
symmetry, 82
synthesis, ix, x, xi, 100, 155, 221
synthetic rubbers, 155

T

TBP, 161, 163, 164, 165
temperature, 1, 32, 33, 64, 130, 135, 149, 163, 167, 168, 172, 176, 177, 178, 183, 204
tetrahydrofurane, 204
thermal decomposition, 1, 84, 87, 128, 134, 135, 137, 138, 139, 140, 142, 143, 144, 160, 183, 190, 191, 214, 223, 225
thermal stability, ix, 1, 2, 142, 145, 220, 221, 223, 224, 225
thermolysis, xi, 1, 13, 103, 129, 130, 131, 133, 135, 137, 138, 140, 141, 142, 143, 144, 145, 146, 153, 159, 164, 169, 178, 179, 181, 182, 188, 193, 212, 214, 215, 223, 224
total energy, 121
toxicity, ix
transformation, 28, 31, 68, 144, 193, 220, 221, 223, 225
transformation product, 144
transformations, ix, x, 111, 158, 172

U

USSR, 2, 230, 231, 232, 233, 234, 235, 236, 237, 238, 239, 241, 242
UV, 75, 159, 160, 161, 162, 166, 167, 169, 170, 171, 175, 180, 182, 189, 195, 238
UV–illumination, 160, 161, 170, 171, 175, 195

V

valence, x, xi, 3, 7, 16, 26, 29, 32, 54, 70, 121, 122, 124, 129, 132, 145, 150, 151, 152, 153, 154, 156, 157, 158, 219, 220, 225
Van der Waals radius, 30, 117

vector, 13, 25
vibration, 4, 5
vinyl monomers, ix
viscosity, 223
visualization, 29, 30, 73
vulcanization, ix, 1, 155

W

water, 51
weak interaction, xi
91

Z

zeroth order, 28